Safety Design Criteria for Industrial Plants

Volume I

Editors

Maurizio Cumo
Antonio Naviglio
Professors
University of Rome
Rome, Italy

CRC Press
Taylor & Francis Group
Boca Raton London New York

CRC Press is an imprint of the
Taylor & Francis Group, an **informa** business

Library of Congress Cataloging-in-Publication Data

Safety design criteria for industrial plants / editors, Maurizio Cumo,
  Antonio Naviglio.
    p.   cm.
  Includes bibliographies and index.
  ISBN 0-8493-6383-7 (v. 1).   ISBN 0-8493-6384-5 (v. 2)
   1. Industrial Safety.   I. Cumo, Maurizio.   II. Naviglio, Antonio.
T55.S215 1989
670.42′028′9—dc19   88-19123
                                                                  CIP

Direct all inquiries to CRC Press, Inc., 2000 Corporate Blvd., N.W., Boca Raton, Florida, 33431.

© 1989 by CRC Press, Inc.

International Standard Book Number 0-8493-6383-7 (v.1)
International Standard Book Number 0-8493-6384-5 (v.2)

Library of Congress Number 88-19123
Printed in the United States

DEDICATION

To my son Fabrizio

Maurizio Cumo

To my parents Anna Maria and Luigi Eros

Antonio Naviglio

PREFACE

The aim of this book is to provide a picture of safety-related aspects of the design, operation, and maintenance of a high-risk plant or — more generally — of a plant where substances potentially harmful for human health are handled.

The book is addressed to people looking for an introduction to safety design criteria of industrial plants, namely students, engineers, and technicians involved in the design, operation, or maintenance of industrial plants, people working for local or national authorities or agencies, with a responsibility in licensing or control of industrial plants.

The book is not an exhaustive handbook including all the aspects and methodologies employed for the design of industrial plants using hazardous matter, but it is an introductory guide addressing the reader to the most relevant items and aspects that should be known when operating, designing, and licensing this kind of plant.

The most relevant concepts and methodologies are introduced and a continuous reference to specialized books or papers dealing with the specific themes is done to allow a deeper insight for people looking for a better knowledge.

A trial has been done for a unique treatment for all kinds of medium and high-risk plants; explicit references to the specific kinds of plants are used whenever necessary.

The wide range of topics, including general safety design criteria, risk analysis methodologies, the main damage causes for the plants, the effects of dispersion of pollutants into the environment, and the human health effects of most relevant pollutants is aimed at giving a picture of the various aspects that could or should be considered in the safety or risk assessment of an industrial plant.

The description of the various external and internal events that could prove as initiating events of accidents does not imply that all of them must be considered in the design of a plant; what is important is to know their existence, while the applicability of the single design criterion will be judged by the designer, also on the basis of the rules and regulations locally in force.

The subject has been subdivided into six sections.

The first, dedicated to the concept and definition of industrial risks, is aimed at introducing the concept of the technological risk (Chapter 1) and at the quantitative indentification of the risk for human health, relative to the main high-risk substances (chemical, radiological) (Chapter 2).

Section II, dedicated to the main aspects of risk analyses of industrial plants, includes the definitions of the main magnitudes commonly employed in risk analyses and assessment (Chapter 3), the main phases of such analyses and assessment (Chapter 4), and the description of methodologies for a system reliability assessment (Chapter 5).

The third section, dedicated to the main design criteria of a general validity and to the causes of damage for the components of an industrial plant, internal to the process, includes a description of general safety design criteria (Chapter 6), the analysis of causes of damage potentially responsible for a release of hazardous substances of thermodynamic nature (Chapter 7) and chemical nature (Chapter 8), and the analysis of the relevance and of the role of instrumentation and control for the operation and protection of risky plants and for the timely monitoring of hazardous matters releases (Chapter 9).

The fourth section assumes that damage to a plant might happen and analyzes the dispersion mechanisms of pollutants into the environment, with the aim of identifying methods to limit the exposure to risk for population and operation personnel. It includes a description of methods for the evaluation of the consequences of effluents dispersion (Chapter 10), methods to limit exposure during normal operation of plants (Chapter 11), and methods to limit exposure following accidents (Chapter 12).

Section I in Volume II, dedicated to the potential causes of damage for the components of an industrial plant external to the process, includes the analysis and criteria to defend the plant against natural events, comprehending the soil-structure interactions (Chapter 1), flooding (Chapter 2), the analysis and design criteria against man-induced events, comprehending fire (Chapter 3), clouds explosions, toxic clouds, and sabotage actions (Chapter 4), and area events (Chapter 5); an insight is performed on the role of human factors for the safety of plants (Chapter 6).

The first five sections are especially addressed to plants handling high-risk substances and to associated criteria to limit the entity of the health risk; Section II in Volume II, on the contrary, is focused on "conventional risks" in industrial plants, which are by far mainly responsible for injuries to personnel involved in operation and maintenance of industrial plants. This section includes an analysis of general industrial risk-protection criteria and a description of most common health risks for personnel operating in industrial plants.

The relevance of the analyzed topics, together with their multidisciplinarity, would justify a very deep analysis; they are only introduced in this book, whose aim is to give a general feeling on the variety and complexity of aspects to be considered, so as to allow the most appropriate decisions in view of further research. Nevertheless, the modern approach to safety is of a global nature, considering and comparing the different points of view to build a system assessment which needs specific contributions of different nature and sciences in harmonic construction. In this sense the book may offer a method and a sensibility for the new mind which is necessary to master the safety of high-risk installations.

Maurizio Cumo
Antonio Naviglio
University of Rome
May 31, 1987

THE EDITORS

Maurizio Cumo is Professor of Nuclear Plants at the University La Sapienza of Rome, where he is also director of the postgraduate School for Nuclear Safety and Radioprotection.

He received his doctorate in Nuclear Engineering at the Politechnic University of Milan in 1962. Since that year, without discontinuities, he has participated in much thermohydraulic research at the ENEA Research Center Casaccia near Rome as well as at the University of Rome. In this field he has authored or coauthored more than 150 scientific publications both international and national as well as 2 books. A member of the Assembly for International Heat Transfer Conferences, the Executive Committee of the International Center for Heat and Mass Transfer, the EUROTHERM Committee, and the European Two-Phase Flow Group with particular attention to experimental developments, presently he is acting as President of the Italian Commission for Nuclear Safety and Health Protection and as a member of the board of the Italian State Agency for Nuclear and Alternative Energies (ENEA).

He is also Chairman of the Italian Association of Nuclear Engineering (ANDIN), Vice Chairman of the Italian Society of Standards (UNI), a member of the board of directors of the International Solar Energy Society (ISES), the American Nuclear Society, the American Institute of Chemical Engineers, and the New York Academy of Sciences.

Biographical references are provided by Who's Who in the World, Who's Who in Europe, International Book of Honor, International Who's Who in Education, Dictionary of International Biography, World Nuclear/World Energy Directory, and the International Directory of Distinguished Leadership.

Antonio Naviglio is Professor of Thermal Hydraulics in the Department of Energetics, Faculty of Engineering, University of Rome.

He received the Italian "Laurea" in Nuclear Engineering in 1973 from the University of Rome. During 1973 to 1975 he worked as a process engineer in a major Italian engineering company. In 1975 and 1976 he worked for the Italian Agency for Nuclear Safety and Radiological Protection (ENEA-DISP). During 1976 to 1981 he worked as a process engineer mainly in the field of thermal hydraulics, for the Italian Electric Power Authority (ENEL). Since 1981, he has been working at the University of Rome, first as assistant professor and then as Professor of Thermal Hydraulics.

Professor Naviglio is an expert member of the Italian Committee for Nuclear Safety and Radiological Protection, the executive committee of ANDIN, the Italian Association of Nuclear Engineers and of ANIAI, the Italian Association of Architects and Engineers, and the Director of UNITAR/UNDP Centre on Small Energy Resources in Rome.

Professor Naviglio has authored or coauthored some 90 scientific publications in the field of heat transfer and energy exploitation.

The research activity of Professor Naviglio has been mainly devoted to heat transfer phenomena, to complex thermal hydraulic phenomenologies affecting the performance and safety of equipment both for nuclear and for chemical plants, and to the development of innovative processes allowing energy saving and minimizing environmental impact.

CONTRIBUTORS

Luciano Bramati
ENEL - DCO
Rome, Italy

Stefano Clementel
Milan, Italy

Remo Galvagni
Milan, Italy

Augusto Gandini
Dipartimento Reattori Veloci
ENEA - CRE
Rome, Italy

Giovanni Lelli
Dipartimento Reattori Veloci
ENEA-CRE Casaccia
Rome, Italy

Antonio Moccaldi
ISPESL
Rome, Italy

Sebastiano Serra
Rome, Italy

Giancarlo Tenaglia
Project Manager
ENEA
Rome, Italy

Giuseppe Volta
System Engineering Division
CEC Joint Research Center
Ispra, Italy

Fausto Zambardi
Instrument Systems Engineer
ENEA
Rome, Italy

TABLE OF CONTENTS

Volume I

Volume II

Section I
The Risk

INTRODUCTION

The concept of risk is introduced and the distinction between the technological risk and the natural one is explained. The technological risk may regard several aspects, among them the damage to a plant, its unavailability, and the noxious impact for the environment or for human health.

In order to perform a risk analysis and to provide the fundamental elements able to guide a decision about an industrial plant (to operate or not, to stop it for maintenance or not, to license it or not, etc.), it is important to well identify and measure the overall risks contracted with the two options (to do the action, not to do it). The risks evaluation involves economical, occupational aspects that are typical of the specific case to be examined: there also may be aspects relating to human health. To contribute in the risk evaluation under this point of view, i.e., certainly the most complex one, a trial has been done to identify the possible risky substances for human health and to associate to each of them objective parameters (concentration in air, in water, etc.) which are representative of well-specified consequences for man. The knowledge of maximum admissible concentrations of the various substances and of the connected health consequences is, also, a starting point in the design, operation, and maintenance of plants handling such substances as in the design of the various systems, equipment, and components of the plants.

Chapter 1

THE TECHNOLOGICAL RISK

Giuseppe Volta

TABLE OF CONTENTS

I. THE CONCEPT OF RISK

The word risk has been always associated with the concepts of uncertainty, possibility, and loss or damage.

In the history of economics and in particular of insurance economy, starting from the 18th century, the term risk has acquired the scientific meaning of mathematical expectation or expected value of the monetary equivalent of damage, assumed as a random variable. This quantified risk, taken in consideration by economists and insurers, embodies uncertainty through a probability distribution, supposedly derived by statistical inference from the experience of a significant number of events. To underline this implicit hypothesis, some authors propose the use of the word "risk" to address a situation for which the statistical inference is possible and the use of the word "uncertainty" to address those situations for which such inference is not possible.[1] However, still in the domain of insurance economics, the distinction between the two types of uncertainty has been contested by De Finetti.[2] In accordance with his subjectivistic concept of probability, all types of uncertainty, even those regarding a unique event, can be measured by a probability.

Independently from the dispute about the applicability of probability as measure of some types of uncertainty, insurance economy has founded a theory of risk firmly based on a monetary equivalent of damage and on a probabilistic measure of uncertainty.

In connection with the most recent dramatic industrial development, risk is becoming widely used outside of the traditional frontiers of economics to designate the potential negative side effects of technology. The new meaning of the term corresponds to a complex function of a set of events, of their likelihood, and of their monetary and nonmonetary consequences. The complexity of this function corresponds to the complexity of the reality that it represents: various types and levels of uncertainties, multiple consequences distributed in time and space, and inhomogeneous and continuously varying reference values.

The new meaning of risk is more near to the philosophical "negative aspect of possibility" than to the simple mathematical concept developed by the insurance economy. The technological risk can be in fact defined as the negative aspect of the possibilities offered man by technology.

II. THE DEEP ROOTS OF TECHNOLOGICAL RISK

Before considering detailed aspects of technological risk let us put forward some considerations about the reasons why technological development yields risk.

Technologies are produced with the aim of satisfying actual or potential needs. Therefore, technologies build a world of "artificial" possibilities that is superimposed onto the world of "natural" possibilities. As in the world of natural possibilities there is honey and hemlock, so in our artificial world there is hexaclorophene, a useful disinfectant, and dioxin, a byproduct of its production. The problem for humans is that of selecting the positive possibilities and rejecting the negative ones.

In nature a capacity for discernment is grown through a secular learning process that allows us to select easily honey from hemlock. In the artificial world the discernment requires a prevision about realities never experienced. This prevision is made difficult by various factors.

The first factor is the complexity of the world created by technology. Complexity means uncertainty and in this sense one can say that industrial development is creator of uncertainty.

The second factor is the relational, systemic character of the negative aspects of technologies. "Prevision" must always take into account the context in which a possibility can be realized. But where are the boundaries of this context? The dispute about the alternative "standards of emission" vs. "standards of environmental impact" illustrates well the complexity of this factor.

The third factor is the increased vulnerability, i.e., the increased potential instability of sociopolitical systems in front of accidents of technological origin.[3] The Seveso accident, for instance, shook the Italian political system much more than the great earthquakes of Friuli and Campania which occurred in the same years. As every scientist knows, the study of intrinsically nonlinear instability phenomena is always more difficult than the study of any linear resistance problem.

The fourth factor is the fast evolution of the technological system, which does not give time for the knowledge of side effects. Moreover the research on abnormal and noxious behavior of artificial systems, i.e., the research on technological risks, has, in the present cultural context, a prestige and a priority much lower than the research on productive aspects.

III. TECHNOLOGICAL RISK AND NATURAL RISK

We mean as a natural risk the possibility of damage not caused by a free choice of humans. A typical example is the seismic or metereological risk. The frontier separating the natural from the technological risk is fairly fuzzy. Human activities have conditioned so heavily natural equilibria that many of the so-called natural risks have also some remote human origin.[4] However, technological risk is directly characterized by a human decision at its origin.

As risk stems from a decision, it has always a motivation of some benefit for someone. Technological risk is one side of a two-sided (good and bad) coin. Moreover, it is a side of a coin that humans in many cases are obliged to spend because humans cannot avoid to act, to progress, and by consequence to decide, to choose. The decision of not doing something is also a choice. For instance, man cannot consider the risk at work, without considering the risk of unemployment.

A society can in some way determine the profile of the mortality curve (number of deaths for age intervals), but cannot avoid having a mortality curve.

IV. CATEGORIES OF TECHNOLOGICAL RISKS

We can distinguish various categories of risk regarding the time of the noxious event, the time of damage, the subjects at risk, and the attitude of subjects facing risk.

The noxious event can be continuous or frequently repeated in time, or it can be occasional and rare. The first case corresponds to routine risk. A typical example is the risk associated with environmental pollution. The second case corresponds to nonroutine risk, accidents, and in particular rare but catastrophic accidents.

With reference to time, the damage of a noxious event can be revealed promptly or later. Therefore, we can have the category of prompt risks and that of delayed risks. An example of the latter is the teratogenic and the mutagenic effect of some chemicals and of radiation.

Damage can be "deterministic" or "stochastic". The distinction has a particular importance for the risk associated with low exposure to harmful agents. The deterministic damages are those for which the causal relationship between an event and its effect can be ascertained. Stochastic damages are those for which that casual relationship cannot be ascertained for individual cases, but can be ascertained statistically on a population. Damages due to low levels of exposure to chemicals or radiation are stochastic in nature.

Regarding the subject exposed to risk, we can distinguish individual and collective risks. Often some confusion is made between individual risk in the true sense and expected value of a risk spread over a population divided by the number of people. It should be more appropriate to call the first individual risk, and the second average individual risk.

Regarding the attitude of people facing risk, an important distinction is between voluntary and involuntary risk. The distinction is important because it is connected with the individual

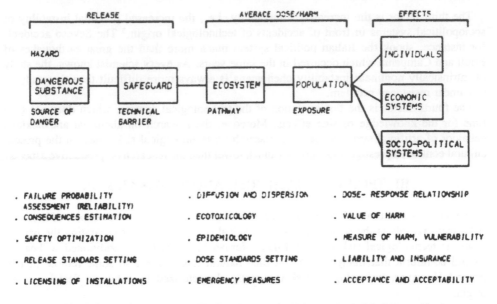

FIGURE 1.   Technological risk levels and some related assessment and decision problems.

acceptability of risk. As at the origin of technological risk there is always a choice, a decision, we can recognize the voluntary or the nonvoluntary character of risk to the degree of participation of the subject to that decision.

V. RISK ANALYSIS

Risk analysis provides a formal representation of the possibilities of damage connected with the operation of a system.[5,6] The analyses can have as the object various categories of risk and various boundaries of the system. Moreover the analysis is always oriented in some way to a decision. Figure 1 outlines the characteristic levels of a risk analysis and the connected decisional contexts. Figure 2 outlines the process of analysis.

The aim of risk analysis is a formal representation, i.e., a qualitative model that can become quantitative to the extent that some measures for the variables and the parameters of the model can be defined. In this sense risk analysis and safety analysis are synonymous. However, in common language, the term safety analysis is restricted to signify the first level of Figure 1, i.e., the part of the analysis that concerns what happens inside the confines of a plant or inside the physical boundaries of a process.

VI. MEASURE OF DAMAGE

The most fundamental problem in risk analysis and that most overlooked is that of measure.

As is well known, given an empirical reality, we can represent some of its characteristics by formal nonnumerical models, e.g., graphs, or by formal numerical models, e.g., equations. The problem of measure is that of finding a rule that allows the use of numbers to represent the properties of the empirical reality.[7]

Risk is a possibility of damage, so it raises the problem of measuring very different things: "damage" and "possibility".

The measure of damage could seem a simple problem. However, it merits some clarification. The simplest solution, that makes easy the comparison of different damages, is to find a monetary equivalent of damage. Insurers, in a market economy, have the role of

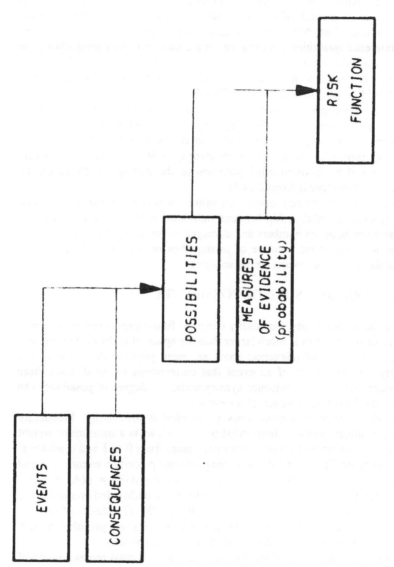

FIGURE 2.   Steps in risk analysis.

finding this equivalent. However, not all damages are and/or can be covered by insurance. Damages that could be in principle monetized for their intrinsic characteristic can raise insurmountable problems when they are delayed in time. The use of a discount rate, the solution offered by economic sciences for taking into account time, is illusory and misleading when the time span involved is more than some decades.

When monetary equivalent is not acceptable, the measure of damage can be approached by the use of countable entities: deaths, injuries, working hours lost, etc. The absolute measure of these entities is, in general, of limited interest. Relative measures are much more interesting, e.g., deaths/Km, death/Mwh, etc. Using relative measures, one must however pay attention to the reference quantities. for instance, in the case of energy production, one can formulate the damage with reference to the energy offered to users, or to the energy transformed, or power, etc. Numbers change and the results of comparisons can also change.

Special attention should be paid to the distinction between incremental values of a given damage and gross values. Damages to the public due to environmental pollution are incremental because they are added to damages suffered by the public from other origins. Damages due to accidents at work place are gross because they cannot be added to the damages of being unemployed. Therefore, it is not correct, in comparing nuclear vs. fossil fuel energy, to sum the damages caused by environmental pollution to the damages suffered by the workers employed in the respective industrial cycle.

There are then cases in which one can assign only ordinal measures to damages, that is, one can establish only an order relationship between individual damages. In this case one can adopt a correspondence between numbers and damages in the sense that numbers have property of order (let us think of the measure of earthquakes on the Mercalli scale), but statistics on these numbers, such as mean, are meaningless.

VII. MEASURE OF POSSIBILITY

The second term of risk is possibility. Possibility concerns future events that may or may not happen. The space of possibilities is much larger than the space of realities. Possibilities do not all have the same degree of relevance; some are more possible than others. The "degree of possibility" is the quality of an event that corresponds to the distance from certainty. If we consider certainty and evidence synonymous, the degree of possibility can also be considered as the degree of evidence of an event.

The measure of evidence most commonly known and used is probability. Probability (independently from any subjectivistic or frequentist point of view) is a measure or weight of the evidence that a given event will occur. It assumes values from 0 to 1 and is additive. This measure that we indicate "p" obeys the following axioms: p (certain event) $= 1$ and p (impossible event) $= 0$. If A and B are nonconnected events: $p(A \cup B) = p(A) + p(B)$.

The additive probability as a measure of uncertainty enjoyed an undisputed predominance in the fields of science, technology, and economics until the 1970s. The theory of risk has been founded on this type of measure, to the point that risk analysis and probabilistic risk analysis or probabilistic risk assessment (PRA) have become synonymous.

In more recent years there has been a flourishing of studies on formal representation of human knowledge. The interest for these studies corresponds to a practical need, originated by the development of information technologies: the need of simulating human behavior in order to anticipate human actions or to replace men by adequately programmed computers. Knowledge representation implies also representation of the uncertainty which imbibes that knowledge.

In this way a critical reappraisal of measures of uncertainty has recently been started, and attention has been focused on the assumptions which are behind additive probability measures. One can appreciate the importance of the additivity axiom realizing that it implies that where he assigns the measure p to the evidence of an event he must also assign the measure 1-p to the totality of the other possible events. In cases of frequent and precise events, perceived as cases of a class of similar events, the measure of uncertainty corresponds to a repeated or repeatable experience. Additivity for this measure is self-imposing as a condition of coherency in human judgment. In fact, in a class of frequent events, the singular case is perceived jointly with the complementary cases. However, in case of unrepeatable or imprecise events, the judgment on those events does not imply necessarily a judgment on complementary events.

To take into account the situation of rare and/or imprecise events, nonadditive measures are being considered. Among these measures, those that are generally called fuzzy measures, an important formal category is represented by the "plausibility" measure "g" which includes, as a special case, the additive probability. Below we compare the axioms and the composition rules of p and g. If ϕ is the impossible event and and ω is the certain event,

$$p(\phi) = 0 \qquad g(\phi) = 0$$

$$p(\Omega) = 1 \qquad g(\Omega) = 1$$

$$\text{and if } A \cap B = \phi$$

$$p(A \cup B) = p(A) + p(B) \qquad g(A \cup B) \geq \max\{g(A), g(B)\}$$

If the inequality becomes equality, plausibility becomes, according to the terminology proposed by Zadeh[8] and then by Dubois and Prade,[9] possibility.

The use of the max operator instead of the ordinary addition in the third axiom should not cause surprise when we are dealing with events having a particular character. This character is reflected in the fact that the subject who assesses the uncertainty or degree of evidence judges in a singular mode, i.e., leaving out of consideration the uncertainty of all other events. It is the case of rare events known in an imprecise mode. Human judgment, when dealing with such types of events, tends to adopt a threshold logic instead of an additive logic.

The sectors of risk analysis for which nonadditive uncertainty measures show a special interest are those in which infrequent and complex events, reducible with difficulty and imprecision to elementary components, are involved.

VIII. DECISION

The risk analysis of a technological system as outlined in Figure 2 leads to a risk function passing through two successive steps.

In the first step the binary relation that links events to harmful consequences is identified. This relation produces the space of possibilities.

In the second step one assigns to each possibility a measure of uncertainty.

These two steps imply simplifications and hypotheses which are already, by themselves, decisions. But the decision that we intend to consider here is the use of the risk function.

In the most simple case, risk function is given by a probability distribution of a consequence (damage), discrete in case of a discrete number of events, continuous when a continuous random variable is associated with the events. This function can be represented in various ways. In the history of risk analysis two most common types of representation are those given in Figures 3 and 4.

The decision problem is comparing two risk functions or comparing a risk function with some acceptability criteria.

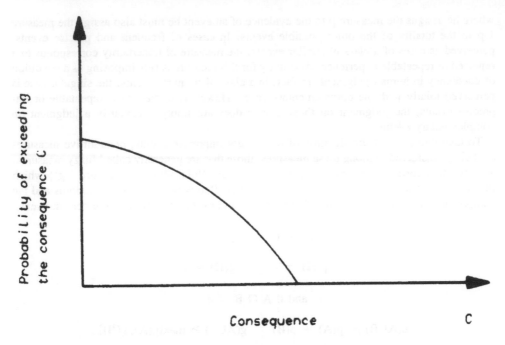

FIGURE 3. Risk function — complementary cumulative distribution function of C.

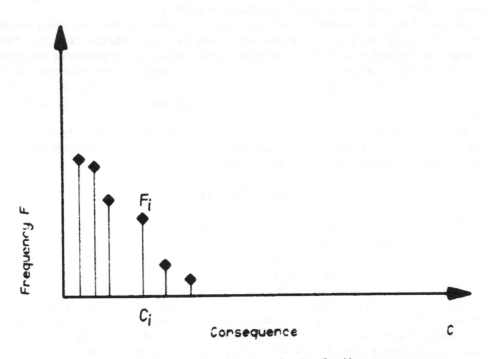

FIGURE 4. Risk function — discrete frequency function of accidents consequences.

The most widely recognized and diffuse decision theory is that going under the name of "expected utility". This theory is a theory of individual decisions, born in the 17th century, but particularly developed and formalized in this century by De Finetti, Ramsey, Savage, and von Neumann. According to this theory, if some axioms are respected, and if we assume that the measure of the consequence of an event is identical to the measure of its utility, the distribution of a consequence, i.e., a risk curve, from a decision point of view, is equivalent to its expected value. The definition of risk as "sum of the probability of harmful events multiplied by their consequences", implies adhesion to this theory.

In reality it has been recognized that human behavior does not correspond to this theory. This fact should not be imputed to "lack of rationality" of the decision maker, but rather to the lack of correspondence of the axioms on which the theory is based to the reality. This theory is inadequate mainly for high levels of decision, e.g., those required by potentially catastrophic events.[10]

The limits of the expected utility theory are connected to the limits of probability as a measure of uncertainty.

Probability and expected utility are, in fact, both models adequate for sets of frequent events, repeated, having easily identifiable consequences, events for which the temporal sequence could be considered irrelevant and in this sense perfectly exchangeable. They are models adequate for realities which are conceptually similar to the situation of games of chance, but technological risk can only partially be brought back to that conceptual framework.

In the practice of administrative decisions concerning safety and risk, the most common approach is that of judging events in the singular mode. This leads to the use of a threshold logic considering each possible event as the totality. This approach which privileges the extremes in comparison with the average values, and that which seems more sensible when dealing with major hazards, has, however, not yet found a theoretical development equivalent to that of the classical decision theory.

REFERENCES

1. **Knight, F. H.,** *Risk, Uncertainty and Profit,* Houghton-Mifflin, Boston, 1921.
2. **De Finetti, B. and Emanuelli, F.,** *Economia Delle Assicurazioni,* UTET, Torino, Italy, 1967.
3. **Lagadec, P.,** *Major Technological Risk,* Pergamon Press, Oxford, 1982.
4. **V.V.A.A.,** *Development and Environment,* Mouton, Paris, 1971.
5. **Apostolakis, G., Garribba, S., and Volta, G.,** *Synthesis and Analysis Methods for Safety and Reliability Studies,* Plenum Press, New York, 1980.
6. **Volta, G.,** Review of methods for risk analysis, *Angew. Systemanal.,* 2(2), 158, 1981.
7. **Kranz, D. H., Luce, R., Suppes, P., and Tversky, A.,** *Foundations of Measurement,* Vol. 1, Academic Press, New York, 1971.
8. **Zadeh, L. A.,** Fuzzy sets as a basis for a theory of possibility, *Fuzzy Sets Syst.,* 1, 3, 1978.
9. **Dubois, D. and Prade, H.,** *Theorie des Possibilities,* Masson, Paris, 1985.
10. **Volta, G. and Otway, H.,** The logic of probabilistic risk assessment versus decision levels, *Nucl. Eng. Des.,* 93, 2 and 3, 329, 1986.

Chapter 2

HIGH RISK PLANTS AND RISK IDENTIFICATION

Sergio Paribelli and Antonio Moccaldi

TABLE OF CONTENTS

I. INTRODUCTION

The knowledge of the risk associated with the introduction of chemical or radioactive substances into the environment constitutes the essential presupposition for any planning, construction, operation, or maintenance activity of plants which deal with such substances.

Such knowledge includes both the effects of these substances upon men, animals, and plant life and the consequences of phenomena of dispersion, through which to correlate quantitatively the entity of releases in the plant to values of concentration in the air, in the water, and on the ground.

There are international, national, and local regulations which furnish the maximum allowable values of concentration of the various substances to avoid injurious effects upon man. These obviously constitute the principal reference in the planning, construction, and operation of plants. Attention must be paid to the fact that the quantitative risk evaluation associated with a chemical or nuclear plant requires, behind biological and medical knowledge, all chemicophysical, toxicological, radiological, and technological characteristics of all the substances stored, handled, or obtained as intermediate or final products of the various processes in the plant.

In this chapter the main qualitative and quantitative elements which allow the identification of the damage to health, associated with the principal chemical substances and with substances of radiological risk, are defined. In particular, the exposure limits and the biological exposure limits for workers and for the population are introduced, and a list is given of those substances considered occupationally oncogenic.

Finally, plants and chemical substances defined as being of high risk European recommendation* on the major accident hazards of chemical industrial activities, are identified.

II. CHEMICAL PLANTS

A. Exposure Limits for Workers

Studies on the maximum allowable concentrations (MAC) began before the last world war, have been successively taken up again and extended in many industrialized countries, in particular in the U.S. and in the U.S.S.R., and have led to the compilation of continually up-dated lists of tolerable exposure limits for contaminating noxious substances (smoke, gases, dust).

At first, the concentration limit was understood as the value not to be exceeded even for short periods of time. For preventional and work safety aims, it is of greater importance to make reference to concentration limits integrated in time. This is the actual concern of the American Conference of Governmental Industrial Hygienists (ACGIH) which has substituted the MAC with the threshold limit value (TLV), which refers to concentrations variable in time, with peaks above the average value, compensated by lower concentrations.

The following definitions have been introduced:

- The TLV-time weighted average (TWA) is the average concentration weighted in time, relative to a normal working day of 8 h and a working week of 40 h.
- The TLV-short-term exposure limit (STEL) is defined as a 15-min time weighted exposure which should not be exceeded at any time during a work day even if the 8-h TWA is within the YWA. Exposure at the TLV-STEL should not be repeated more than four times per day, with at least 60 min between successive exposures at the STEL.
- TLV-ceiling limit (CL) is the concentration that should not be exceeded during any part of the working exposure.

---

* In 1982 the European Community Commission issued a directive (ECC/501/24.6.82) concerning major hazards plants.

The notation "skin" following the TLV indicates that even though the air concentration may be below the limit value, significant additional exposure to the skin may be dangerous.

The units of measurement used are the ppm (parts per million) or the mg/m^3. The conversion formula is

$$mg/m^3 = ppm/24.45 \times M$$

where M is the molecular weight of the substance; 24.45 is the volume occupied by 1 g molecular weight of gas at 25°C and 760 mm Hg, in liters.

These are some of the most important abbreviations of the organizations that emit standards and recommendations.

OSHA — The limits recommended by the ACGIH are generally translated in the regulations of the Occupational Safety and Health Administration (OSHA) in the U.S. This agency is a part of the Federal Department of Labor. It has, as its basic responsibilities, the promulgation and enforcement of health and safety standards.

NIOSH — The National Institute for Occupational Safety and Health (NIOSH) is a part of the U.S. Department of Health, Education, and Welfare. This agency carries out research for the development of new standards and the provision for training of personnel.

EPA — The Environmental Protection Agency is referred to "worker Protection Standards for Agricultural Pesticides".

For substances used in agriculture, Food and Agricultural Organisation of the United Nations (FAO) defines the values and limits of reference. These values can be found in the *Codex Maximum Limits for Pesticide Residues*, published by the Codex Alimentarius Commission.

The meanings of the principal notations used are reported in the following:

1. The pesticide residue is any specified substance in food, agricultural commodities, or animal feed resulting from the use of a pesticide. The term includes any derivative of a pesticide, such as conversion products, metabolite reaction products, and impurities considered to be of toxicological significance.
2. The maximum residue limit (MRL) is the maximum concentration for a pesticide residue resulting from the use of a pesticide, according to good agricultural practice, that is recommended by the Codex Alimentarius Commission to be legally permitted or recognized as acceptable in or on a food, agricultural commodity, or animal feed. The concentration is expressed in milligrams of pesticide residue per kilogram of the commodity.
3. The guideline level is used to assist authorities in determining the maximum concentration of a pesticide residue where an acceptable daily intake (ADI) or a temporary acceptable daily intake for the pesticide has not been estimated or has been withdrawn by the joint FAO/WHO Meeting on Pesticide Residues. The concentration is expressed in milligrams of pesticide residue per kilogram of the commodity.
4. The ADI of a chemical is the daily intake which, during an entire lifetime, appears to be without appreciable risk to the health of the consumer on the basis of all the known facts at the time of the evaluation of the chemical by the joint FAO/WHO Meeting on Pesticide Residues. It is expressed in milligrams of the chemical per kilogram of body weight.

The list of toxic substances and relative TLV, MAC, and MRL are in continuous evolution both because of the acquisition of new epidemiological data, which may lead to a variation in the previous limits adopted, and because of the introduction of new substances.

On the other hand, it is important to underline that the exposure limits can assume different values not only according to the level and period of experimentation, but also to the interpretation adopted of the concept of "health", which is essential in all reported definitions. The values recommended by the ACGIH are the result of experimental research on many species of animals (studies regarding the survival from exposure or the appearance of particular modifications of pathological or physiological character) as well as long-term observations of man. The interpretation of the results is based on the concept of no cellular damage.

B. Biological Exposure Limits

The actual dose for a worker exposed to risk is measured according to the biological exposure index (BEI), which is a measure of the amount of a chemical substance absorbed into the body. This parameter is particularly useful in evaluating exposures to substances with significant absorption through the skin

The biological check on the individual is carried out by looking for the substance or one of its metabolites in the organic liquids. The concentrations thus found, if compared to reference values and correlated to environmental concentrations, allow one to evaluate and make decisions about systems of prevention and protection adopted (local exhaust systems, respirators, etc.).

One of the major obstacles to the use of the information derivable from the dose measurements regards the particular metabolism of the different toxic substances; some of these, in fact, are rapidly digested, others need a longer period of time, yet others are of very slow elimination with a tendency to accumulate and oncogenic effect. Such elimination times must be carefully assessed in the study of biological exposure limits.

C. The Effect of Toxic Substances on the Body

The effect of toxic substances on the body depends on many factors including the chemicophysical characteristics of the substances, the way of absorption, the intermediate metabolism, the means of elimination, and the acute and chronic toxicity of the substance itself.

The absorption of toxic substances into the body can happen by means of inhalation, through the digestive system (as in the case of accidental ingestion or inhalation of dust or smoke, or by inadequate observation of the hygiene rules), or finally through the skin, as with water- or lipo-soluble elements or compounds. The substances absorbed are distributed in the different organic tissues according to their degree of water or lipo solubility and undergo a biotransformation for oxidation, reduction, or conjugation. The products resulting from such reactions are eliminated by the respiratory, intestinal, or renal system, or accumulate in particular organs and tissues (body burden) e.g., as lead which substitutes calcium in the bones as a sulfate or bicarbonate. Acute toxicity (brief periods of exposure) is expressed through:

● The lethal dose (LD).
● The LD_{50} is a calculated dose of a substance which is expected to cause the death of 50% of an entire, defined, experimental animal population. It is determined from the exposure to the substance by any route other than inhalation of a significant number from that population.
● The lethal dose low (LDL_0) is the lowest dose (other than LD_{50}) of a substance introduced by any route, other than inhalation, over any given period of time in one or more divided portions, and reported to have caused death in humans or animals.

In the case of absorption through inhalation, the acute toxicity is assessed by means of:

● The lethal concentration (LC).
● The LC_{50} is a calculated concentration of a substance in air, exposure to which for a

Table 1
EVALUATION CRITERIA OF TOXIC
SUBSTANCES (FROM THE NATIONAL
ACADEMY OF SCIENCES CRITERIA)

Toxicity Level	LD_{50}		LC_{50} (ppm)
Extremely toxic	1	mg/kg	0—10
Highly toxic	1—50	mg/kg	10—100
Moderately toxic	50—500	mg/kg	100—1,000
Lightly toxic	0.5—5	g/kg	1,000—10,000
Practically not toxic	5—15	g/kg	>10,000

specified length of time is expected to cause the death of 50% of an entire defined experimental animal population. It is determined from the exposure to the substance of a significant number from that population.

- The lethal concentraction low (LCL_0) is the lowest concentration of a substance in air, other than LC_{50}, which has been reported to have caused death in humans or animals.

In Table 1 we have the criteria of assessment of toxicity of the National Academy of Sciences (U.S.) based on LC_{50} and LD_{50} values.

Chronic toxicity is calculated on the basis of prolonged experimentation with the administration of small daily doses of the substance and is expressed through:

- The toxic dose (TD).
- The toxic dose low (TDL_0) is the lowest dose of a substance introduced by any route, other than inhalation, over any given period of time and reported to produce any toxic effect in humans or to produce carcinogenic, neoplastigenic, or teratogenic effect in animals or humans.
- The toxic concentration (TC) in inhaled air.
- The toxic concentraction low (TCL_0) is the lowest concentration of a substance in air to which humans or animals have been exposed for any given period of time, that has produced any toxic effect in humans or produced a carcinogenic, neoplastigenic, or teratogenic effect in animals or humans.
- The TLm_{96} is defined as the concentration that will kill 50% of the exposed organism within 96 h. Because of the lack of good standardization and the wide variety of species investigated, ratings (ranges of toxicity), rather than a single toxic dose, are generally used to give an indication of the toxicity of substances to aquatic life.

D. Occupational Cancerogenous Substances

An international body, which studies presumed cancerogenous substances and establishes relative regulations is the International Agency for Research on Cancer (IARC) which issues recommendations and information on potential dangers with chemical substances. A list of substances defined as occupationally cancerogenous and deriving from epidemiological studies is given in the following:

- Acrylonitrile
- 4-Aminodiphenil
- Antimony trioxide production
- Arsenic trioxide production
- Asbestos (and smoking)

- Benzene (and leukemia virus)
- Benzidine production
- Cadmium oxide production
- Bis-(chloromethyl) ether
- Chloromethyl methyl ether
- Chromite ore processing
- Coal tar pitch volatiles and particulates
- Coke-oven emissions
- β-Naphthylamine
- Nickel sulfide roasting
- 4-Nitrodiphenyl
- Uranium underground mining
- Vinyl chloride
- Wood dusts

E. Risk Identification: High Risk Substances and Plants

The safety of workers and of the population demands that a preliminary estimation of risk connected with every individual operation and with the functioning of the entire industrial complex is carried out. In fact, the experience acquired shows that malfunctioning, unchecked chemical reactions, leak of toxic substances, etc. cause situations of danger with consequent relevant risk to human life. Through a precise identification of the possible sources of risk, consequent to a systematic evaluation of processes and plants, it is possible to adopt constructive devices and/or active operations to increase the plant safety.

Consequently, not only must great diligence be reserved to active devices for the prevention of the involvement of workers in accidents at work, but also particular attention must be paid to potential risk in the process which can cause fire, explosions, or emission of noxious and/or toxic substances; that is, it is necessary to carry out a systematic examination of the process adopted in such a way as to identify and localize the risks both in normal working conditions and in simple deviations from them, or in accident conditions.

The importance of this analysis is evident if we reflect upon the enormous amount of energy and/or toxic substances involved in many industrial processes. In Europe, the European Economic Community (EEC) (see footnote on page) prescribes "the identification of existing high risks of accident, the adoption of appropriate safety measures, information, training, and equipment necessary for the safety of the people working on site."

Tables 2 to 4, respectively, list the plants and installations which present high risks of accident, according to the above-mentioned EEC directive, the criteria for identification of substances for which it is necessary to take particular safety measures, and, finally, those substances whose presence must be made known to the responsible authority if they are used (manipulated) or stored in quantities superior to those admitted.

For these substances, a file which has been prepared is reported in Chapter 8, in chich the basic chemical and physical characteristics, epidemiological data available, and the TLV values, if these have been fixed, are included. In the above-mentioned chapter the basic elements for the identification of the most important technical characteristics to be foreseen for the plants, with the aim of minimizing the risk, are also given. Each of the reported substances may be identified by the Chemical Abstract Service (CAS) registry number assigned by the American Chemical Society Abstract Service.

III. NUCLEAR PLANTS

A. Ionizing Radiation

Ionizing radiation widely involves man and his environment and contributes to the well-being of man through its wide use in medicine and industry. Ionizing radiation can be of

Table 2
HIGH RISK PLANTS

Installation for the production or processing of organic or inorganic chemicals using for this purpose, in particular
  Alkylation
  Amination by ammonolysis
  Carbonylation
  Condensation
  Dehydrogenation
  Esterification
  Halogenation and manufacture of halogens
  Hydrogenation
  Hydrolysis
  Oxidation
  Polymerization
  Sulfonation
  Desulfurization, manufacture, and transformation of sulfur-containing compounds
  Nitration and manufacture of nitrogen-containing compounds
  Formulation of pesticides and of pharmaceutical products
Installations for the processing of organic and inorganic chemical substances, using for this purpose, in particular
  distillation
  extraction
  solvation
  mixing
Installation for distillation, refining, or other processing of petroleum or petroleum products
Installations for the total or partial disposal of solid or liquid substances by incineration or chemical decomposition
Installations for the production or processing of energy gases, e.g., LPG, LNG, SNG
Installations for the dry distillation of coal or lignite
Installation for the production of metals or nonmetals by a wet process or by means of electrical energy

Table 3
MAJOR HAZARD PLANTS: HIGHLY DANGEROUS SUBSTANCES STORAGE

	Quantities	
Substances or groups of substances	For application of regulation 4	For application of regulations 7 to 12
Acrylonitrile	350	5000
Ammonia	60	600
Ammonium nitrate	500	5000
Chlorite	10	200
Flammable gases as defined in schedule 1 paragraph (c)(i)	50	300
Highly flammable liquids as defined in schedule 1, paragraph (c)(ii)	10,000	100,000
Liquid oxygen	200	2000ᵃ
Sodium chlorate	25	250ᵃ
Sulfur dioxide	20	500

Note: The quantities set out relate to each installation or group of installations belonging to the same manufacturer where the distance between installations is not sufficient to avoid, in foreseeable circumstances, any aggravation of major accident hazards. These quantities apply in any case to each group of installations belonging to the same manufacturer where the distance between the installation is less than 500 m.

ᵃ Where this substance is in a state which gives it properties capable of creating a major accident hazard.

natural or artificial origin and its difference in origin does not determine differences in the radiation itself or in the effects it induces.

By the term ionizing radiation we mean any type of radiation which can produce, directly or indirectly, the ionization of the atoms and molecules of the medium which it crosses. Particles with electrical charge (electrons, protons, α-particles, etc.) whose kinetic energy

Table 4
INDICATIONS FOR IDENTIFICATION OF MAJOR HAZARD SUBSTANCES

Substance	LD_{50} (oral) body weight[a] (mg/kg)	LD_{50} (cutaneous) body weight[b] (mg/kg)	LC_{50} mg 1 inhalation[c]
Very toxic	$LD_{50} \leqslant 5$	$LD_{50} \leqslant 10$	$LC_{50} \leqslant 0.1$
Very toxic[d]	$5 < LD_{50} \leqslant 25$	$10 < LD_{50} \leqslant 50$	$0.1 < LC_{50} \leqslant 0.5$
Other[e]	$25 < LD_{50} \leqslant 200$	$50 < LD_{50} \leqslant 400$	$0.5 < LC_{50} \leqslant 2$
Flammable			
Flammable gases[f]			
Highly flammable liquids[g]			
Flammable liquids[h]			
Explosive[i]			

[a] LD_{50} oral in rats.
[b] LD_{50} cutaneous in rats or rabbits.
[c] LC_{50} by inhalation (4 h) in rats.
[d]   Due to their physical and chemical properties, they are capable of producing major accident hazards similar to those caused by the substance mentioned in the first line.
[e]   Have physical and chemical properties capable of producing major accident hazards.
[f]   Substances which in the gaseous state at normal pressure and mixed with air become flammable and the boiling point of which at normal pressure is 20°C or below.
[g]   Substances which have a flash point lower than 21°C and the boiling point of which at normal pressure is above 20°C.
[h]   Substances which have a flash point lower than 55°C and which remain liquid under pressure, where particular processing conditions, such as high pressure and high temperature, may create major accident hazards.
[i]   Substances which may explode under the effect of flame or which are more sensitive to shocks or friction than dinitrobenzene.

is sufficient to produce ionization by collision, are directly ionizing. Instead, those particles without an electrical charge (neutrons, photons, etc.) which, interacting with matter, can activate ionizing particles or cause nuclear reactions, are indirectly ionizing.

The biological effects of radiation are a function of the energy which is transferred, held, deposited, and absorbed in the living matter of the whole organism or in one of its organs or tissues, in consequence of phenomena of ionization or stimulation which takes place in the atomic or molecular constituents.

B. Source and Dosimetric Measurements

A radiative source is characterized by the number of disintegrations in the unit of time. The activity A of a radiation source is defined as

$$A = dN/dt \tag{1}$$

where dN is the number of spontaneous nuclear transformations which take place in a given quantity of the radionuclide considered, in the interval of time dt; by transformation we mean a change in nuclide or an isometric transition.

The unit of measurement of the activity in SI is the becquerel, Bq:

$$1 \ Bq = 1 \ s^{-1}$$

The old special unit, the curie (Ci), whose numerical value is $1 \ Ci = 3.7 \ 10^{10} s^{-1}$, is still widely used: $1 Bq = 27 \ 10^{-12} \ Ci$.

To obtain information on the radioactivity of a material it is necessary to make recourse to the concept of specific activity As. This measurement, which is expressed in becquerels per gram, represents the number of disintegrations per unit of time which occur in the mass unity of the substance.

The energy E, released by radiation in a given volume is defined as

$$E = R_{in} - R_{out} + \Sigma Q$$

where R_{in} represents the radiant energy incident in the volume considered, that is, the sum of the energies of all the directly or indirectly ionizing particles which enter into the volume considered, R_{out} is the energy leaving the volume, and ΣQ is the sum of all the energies released in the volume, less the energies absorbed for transformations of nuclei of the particles.

The absorbed dose, D, in volume of mass due, is defined as

$$D = d\bar{E}/dm$$

$\bar{E}$ is the average value of the energy transferred. The unit of measurement of the absorbed dose in SI is the gray (Gy):

$$1 \text{ Gy} = 1 \text{ J kg}^{-1}$$

Ionizing radiations cannot be directly detected by the human senses, but they can be detected and measured by a variety of means including photographic films, geiger tubes, and scintillation counters. Measurements made with such detectors can be absorbed by the body or by a particular part of the body. When measurements are not possible, e.g., when a radionuclide is deposited in an internal organ, it is possible to calculate the dose absorbed by that organ if the activity in it is known.

The effective dose equivalent H' is defined as $H' = H \times w$.

The factor w takes into account the sensitivity to the radiation damage of different tissues. The unit of measurement of the dose equivalent is the sievert (Sv):

$$1 \text{ Sv} = 1 \text{ J kg}^{-1}$$

Still in common use in dosimetry is a special unit, the rem: w

$$1 \text{ rem} = 10^{-2} \text{ Sv}$$

Absorbed dose is the quantity of energy imparted by ionizing radiation to a unit mass of matter such as tissue. Absorbed dose is expressed in a unit called gray (1 Gy = 1 J/kg).

Dose equivalent — It is expressed in a unit called sievert and its symbol is Sv. Dose equivalent is equal to the absorbed dose multiplied by a factor that takes into account the way a particular radiation distributes energy in a tissue, thus influencing its effectiveness in causing harm.

For γ-rays, X-rays, and β-particles, the factor is set at 1; therefore, the quantities are numerically equal. For α-particles, the factor is 20, so that 1 Gy of α-radiation corresponds to a dose equivalent of 20 Sv.

The quantity obtained by multiplying the average effective dose equivalent by the number of persons exposed to a given source of radiation is called the collective effective dose equivalent.

C. Sources of Radiation

Radiation can come both from sources outside the organism and from radio nuclides incorporated by inhalation, ingestion, or through wounds; in evaluating the assumed dose for external irradiation, the penetration capability, and the depth and geometry of the various organs and tissues, the geometry and dimensions of the radiative source (puntiform, extended) must be taken into consideration. For internal irradiation, it is important to know the metabolic aspects of the contamination of the body, in particular the ways of introduction and elimination and the accumulation in the body of a radionuclide.

Man is exposed to natural radioactivity, due to several contributions, whose intensity varies greatly with geographical position.

Almost half of the assumed average dose for a man from external irradiation of natural origin derives from cosmic rays; radiation at high energy originating from the sun and from interstellar space (primary cosmic rays), interacts with the atomic nuclei of the elements present in the atmosphere and produces ionizing particles and electromagnetic radiation (secondary cosmic rays). This radiation is constant enough in time at sea level, but is influenced by the geomagnetic latitude and notably increases with altitude through the effect of the magnetic fields of the earth and through the shielding of the air.

A second cause of irradiation for humans is constituted by natural radioactive materials present in the ground (soil and rocks), in particular potassium (40), rubidium (87), and the series of radioactive elements arising from the decay of uranium (238) and thorium (232).

Naturally, the levels of terrestrial radiation differ around the world, as the concentrations of these materials in the crust of the earth vary. An average dose of external irradiation to the population comes from them. This varies from 0.3 to 0.6 m Sv per individual per year, with peak values which go up to 1.4m Sv/year.

Radioactivity in the diet comes from other radionuclides (uranium and thorium series) present in air, food, and water (lead, 210, and polonium, 210). Potassium (40) is taken into the human body in the diet and is the major source of internal irradiation.

The effective dose equivalent from these sources of internal radiation is 370 μSv/year on the average. However, the most important of all sources of natural radiation is an odorless, invisible gas, $7^1/_2$ times heavier than air, called radon. Very high levels of radiation can result especially if a house happens to stand on particularly radioactive ground or has been built with radioactive material (granite, tuff, etc.). When radon enters a house, either from the walls or through the floor, the concentration builds up because the immediate decay products of radon-222 of the limited supply of air from the outdoors are solid daughters, which attach themselves to dust particles in the air. When these are inhaled, they irradiate the lung.

The effective dose equivalent for natural sources taken as a whole varies, depending on the part of the world, from 100 to 400 × 10^{-5} Sv/year, with an average world figure of 200 × 10^{-5} Sv/year. Artificial sources of radiations are

- Medical practices
- Nuclear explosions
- Nuclear energy
- Miscellaneous sources

Medicine is by far the greatest source of human exposure to man-made radiation. Indeed, in many countries it is responsible for virtually the whole dose received from artificial sources.

Radiation is used both for diagnosis and treating diseases. X-ray machines are one of the most widespread and relevant radiation sources; new diagnostic techniques using radioisotopes are spreading rapidly.

The external doses from sources which are of use in medicine are characterized by a high dose rate, which concerns a restricted part of the organism only. Although the use of radiation in medicine offers enormous direct benefit to patients, the fact remains that the average effective dose equivalent per man is 500 μSv/year, almost entirely from diagnostic X-ray procedures.

Further artificial radioactivity spreads throughout the world as a result of nuclear weapons tests in the atmosphere. Radioactivity from the upper atmosphere is transferred slowly to the lower atmosphere and then, more quickly, to earth (fallout). The radionuclides occurring in fallout are inhaled directly or ingested with food and liquids, both processes causing internal exposure of the body. Radionuclides that emit γ-rays when deposited in the soil, cause external irradiation. In the phase subsequent to a nuclear explosion the major contribution to the exposure is due to the nuclides with a short half-life, among which are Zr (95) and Nb (95); over a longer period Sr (90) and Cs (137) are important; these are nuclides which enter into the alimentary chains.

Finally there are people exposed to ionizing radiation as a result of their work in nuclear power plants, in process and quality control, etc. The total dose from radiation of artificial origin is, on average, about 50×10^{-5} Sv/year.

The percentage contribution of each source is on average:

1. Natural radiation

   - 33% radon
   - 16% γ-ray
   - 16% internal
   - 13% cosmic ray

2. Artificial radiation

   - 21% Medical
   - 0.4% Fallout
   - 0.4% Occupational (only for nuclear workers)
   - 0.2% Miscellaneous

D. Effects of Ionizing Radiations

Radiation at high doses can kill cells, damage organs, and cause rapid death; at low doses, it is suspected to trigger off partially understood chains of events which lead to cancer, leukemia, and genetic damage.

The risks connected with the exposure to ionizing radiation of workers and of the general public lie in the somatic and genetic effects it might have. Somatic effects relate to injuries to cells which are concerned with the maintenance of body functions.

The dose-effects relationship in respect of acute effects immediately following exposure is reasonably well understood in most cases; in general, it is possible to specify a minimum dose and a minimum dose rate which will bring about an observable effect. For delayed effects, however, little is understood about this relationship at present.

For certain effects, such as cancer of the skin, or a cataract of the eye, high doses are necessary. In the case of effects such as cancer of bone or lungs, aplastic anemia, and leukemia, it is not known whether a threshold dose exists at all.

Genetic effects relate to injuries to cells in the gonads which are responsible for the propagation of genetic characteristics to subsequent generations. The tissue of the gonads are more radiosensitive. Irradiation of the germ cells may cause mutations which manifest themselves in later generations. Mutations, once having occurred, are permanent. The great majority of observed mutations are deleterious. No conclusive answer is available to the question of whether a threshold dose exists for mutations. Small doses may be cumulative and the end result may not appear until many generations later.

The manifestation of these effects is limited by applying recommendations limiting the exposure of radiation workers and the general public to ionizing radiation. These recommendations are drawn up by the ICRP* and are embodied in international regulations and the national legislation of the individual countries concerned.

E. Principles of Radio Protection: Dose Limits for Individuals

At present, to reduce the doses, three basic criteria are used which can be summarized as follows:

1.   No practice shall be adopted unless its introduction produces a positive net benefit.
2.   All exposures shall be kept as low as reasonably achievable, economic and social factors being taken into account.
3.   The dose equivalent to individuals shall not exceed the limits recommended for the appropriate circumstances by the commission (ICRP).

These criteria apply in full only to the exposure of radiation workers and of the general public to artificial sources (industrial and other practices involving radiation).

Little can be done about the levels of doses from radiation of natural origin and the fallout from weapons tests.

The use of radiation in medicine is conditioned by the extent to which it is necessary to resort to it. This depends on the extent of the diagnostic and therapeutic requirements that are considered in the best interest of the patient who must undergo radiation treatment.

A substantial reduction in the doses, achieved by applying the criteria mentioned above, involves

1.   Acceptance of the radiation effects as a result of the choice made, that is, the advantages this choice brings weighed against the likely consequences if alternative systems were to be used.
2.   Reduction of the doses to levels below which any further reduction would entail inordinately high costs without achieving the expected advantages.
3.   Compliance with the ruling whereby no individuals are to be exposed to unacceptable doses. The effective dose equivalent is 5 mSv/year for members of the general public and 50 mSv/year for radiation workers. These limits must be complied with, regardless of the cost involved.

For any worker this limit is fixed at 500 mSv/year, in any body tissue (with the exception of 150 mSv/year in the eye crystal).

This could also be adopted for individuals in the population, speaking of effects of dose threshold; but for reasons of caution (biological life is longer than working life, checks are not so easy and less diffuse among nonworkers) the commission retains the adoption of a limit equal to one tenth of the above-mentioned value: 50 mSv/year in any body tissue.

In view of the fact that the absorbed dose is in the range of 1 in 10 and the overall risk factor for fatal cancers is about 1 in 80 per sievert (1.25×10^{-2} Sv^{-1}), the average risk of fatal cancer for a radiation worker exposed to the limit dose is therefore 1 in 16,000 per year ($6.25 \times 10^{-5} \times$ year^{-1}).

If we compare this average annual risk of cancer with the average annual risk of fatal accidents in other occupations, we have the following table:

---

*   The International Commission on Radiological Protection (ICRP) is not "governmental" and its structure and activities do not fall within the jurisdiction of the political authorities of the various countries. The commission is, therefore, independent by definition and, as a result, its recommendations do not have the force of law but are explicitly addressed to authorities that are legally empowered to issue laws and regulations on the matter in the various countries or at the international and supernational level.

Coal mining	1 in 4,000 per year	$2.5 \times 10^{-4} \times$ year^{-1}
Construction	1 in 5,000 per year	$2.0 \times 10^{-4} \times$ year^{-1}
Other types of work	1 in 20,000 per year	$5.0 \times 10^{-5} \times$ year^{-1}

With regard to members of the general public, assuming for a group of the population an annual dose of 0.5 mSv, there is an average risk of cancer of 1 in 40,000 in a year ($2.5 \times 10^{-6} \times$ year^{-1}). In the following table are the average annual risks of death from some common causes:

Cause	Risk of death	
Smoking ten cigarettes a day	1 in 200 per year	($5 \times 10^{-3} \times$ year^{-1})
Accidents on the road	1 in 5,000 per year	($2 \times 10^{-4} \times$ year^{-1})
Accidents at home	1 in 10,000 per year	($1 \times 10^{-4} \times$ year^{-1})
Accidents at work	1 in 20,000 per year	($5 \times 10^{-5} \times$ year^{-1})

REFERENCES

1. **Sax, N. I.,** *Dangerous Properties of Industrial Materials.* Van Nostrand Reinhold, New York, 1984.
2. *Toxic and Hazardous Industrial Chemicals Safety Manual for Handling and Disposal with Toxicity and Hazard Data,* International Technical Information Institute, Minato-ku, Tokyo.
3. *Occupational Health Guidelines for Chemical Hazards,* PB83-154609, Parts 1 to 3, U.S. Department of Commerce, National Technical Information Service, 1981.
4. **Plunkett, E. R.,** *Handbook of Industrial Toxicology,* Chemical Publishing, New York, 1976.
5. **della Sanita, I. S.,** *Inventario Nazionale Sostanze Chimiche,* Rome.
6. Codex Alimentarius Commission, Codex Maximum Limits for Pesticide Residues, Joint FAO/WHO Foods Standards Program, Food and Agriculture Organization/World Health Organization.
7. NIOSH, Registry of Toxic Effect of Chemicals Substances.
8. **Melino, C.,** Lineamenti di Igiene del Lavoro.

Section II
Risk Analysis and Assessment Methodologies

INTRODUCTION

Risk analysis is one of the fundamental logic phases in the decisional process of each man. It is performed all the times the necessity of a choice appears and is performed several times each day by each of us. It is performed unconsciously if the entity of the risks associated to the two alternatives of the choice are of a limited importance (e.g., the choice of arriving earlier adopting a higher velocity with the associated risks, instead of arriving later with the associated consequences, but adopting a lower velocity), while it corresponds to a conscious, well-defined, and sometimes codified decisional process, if the risks are of a relevant importance.

That is the case of all design activities of engineered components, structures, and equipment required by a need (eventually, a hypothesized need), but whose construction inevitably involves a risk for somebody, that may also be a health risk.

That is, typically, the case of the construction of a house or of a bridge: the selection of materials and the use of their strength characteristics in the selection of the shapes is, generally, the compromise between two opposite risks: that of an excessive cost and that of a not sufficiently safe realization.

The experience in construction — in most cases — the existence of rules and regulations help in the decision process and avoid an explicit process of risk analysis for the designer.

Nevertheless, if the complexity of the realization is high, if the kind of the realization is new or relatively new, so that sufficient experience is not available and suitable rules or regulations have not been yet developed, there is the need of a specific analysis of the risks able to give the designer the necessary tools for the design and to give the authorities the necessary tools for their decisions. In the design, construction, and operation of plants handling harmful substances and — most of all — of high-risk plants, the need for a comprehensive risk analysis and assessment, to be updated during the life of the plant, is felt more and more.

What is mostly under discussion is the method to be adopted to perform the risk analysis and the criteria on which to base the final decision (acceptability of the risk). While the second aspect is somehow aleatory, depending on the case examined (kind of plant, place of installation, time of installation, social and economic background, etc.), so that a contribution may be given only identifying (or trying to do it) the correlations between the concentrations of the harmful substances and the objective effects for man and the environment (see Section 1 and Section 5), the first aspect is more easily objectifiable.

Even if the problem of selecting a risk analysis methodology universally accepted is far from solution, the methods, procedures, and instruments developed and proposed already allow the solution of specific problems and — most of all — allow to foresee that the goal of a universally accepted methodology (or set of methodologies) can be achieved. The description of risk analysis and risk assessment methodologies today already available would require a quite wider extension with respect to space available in this book; for this reason, only the most relevant concepts have been recalled with the aim of giving, at least, an idea of the phases of a quantitative risk analysis and a feeling of the power of such an evaluation tool.

Chapter 3

ELEMENTS OF STATISTIC AND EVENT ANALYSIS

Augusto Gandini

TABLE OF CONTENTS

I. PROBABILITY: FUNDAMENTAL CONCEPTS

Real events can be generally associated with one of two main categories: the category of events that can be replicated (by this implying identical conditions) and the category of events that are unique, or rare (in the time scale considered). As examples of the first category events we can mention an outcome (e.g., the number 7) in a roulette round, or the failure of a mechanical component during testing at controlled, reproducible conditions. As examples of the second category events, we can mention the position of a given horse in a horse race, the breakdown of a mechanical system during its mission, or the occurring of an earthquake in a given region and period of time.

If the probability of a given event is a quantity by which the degree of belief, or confidence, we have on its outcome can be measured, based on our knowledge at a certain moment, two distinct definitions can be given, depending on which one of the two above categories such event is associated with. Therefore, with respect to the first category events, we shall have the "frequentist" definition by which, given an event X, the probability, $P(X)$, of its outcome at each trial is given by the ratio between the number of favorable events (i.e., those corresponding to event X) and the number of all possible ones, assuming identical conditions at each trial (as is generally the case, e.g., during a roulette round). With respect, instead, of the second category events, we shall have the "subjective" definition by which, given an event X, the probability, $P(X)$, of its outcome corresponds to a personal, (i.e., subjective) quantified assessment of its likelihood (e.g., that a given horse, on which we bet, will win the first position in a horse-race). This assessment is assumed satisfying the criteria of self-consistency and compatibility with the available information.

The joint outcome, or intersection, of two generic events X and Y is indicated with the notation $X \cap Y$ (X and Y). The probability of the outcome of this joint event, $P(X \cap Y)$, is usually denoted as $P(XY)$. Introducing the concept of conditional probability $P(X|Y)$, i.e., the probability of the outcome of X given Y, the following relationship can be written:

$$P(X \cap Y) \equiv P(XY) = P(X|Y)P(Y) \tag{1}$$

The extension to multiple events is immediate. For example, in the case of three events X, Y, and Z, it results

$$P(X \cap Y \cap Z) \equiv P(XYZ) = P(X|YZ)P(Y|Z)P(Z) \tag{2}$$

If, beside X, we introduce its complement $\overline{X}$ *(not* X), i.e., the event that X does not occur, these two events excluding each other, the probability of their joint outcome will result null, i.e.,

$$P(X\overline{X}) = P(X|\overline{X})P(\overline{X}) = 0 \tag{3}$$

Let us consider then the disjoint outcome of two generic events X and Y. We call this event the union of X and Y and denote it with $X \cup Y$ (X or Y). The probability of the outcome of this event, $P(X \cup Y)$, is usually denoted as $P(X + Y)$. The following expression can be written:

$$P(X + Y) = P(X) + P(Y) - P(XY) \tag{4}$$

where $P(XY)$ corresponds to the probability of the joint outcome (intersection) of events X and Y. Also for this case the extension to multiple events is immediate. For example, in the case of three events X, Y, and Z, it results

$$P(X \cup Y \cup Z) \equiv P(X + Y + Z) = P(X) + P(Y) + P(Z) - P(XY)$$
$$-P(XZ) - P(YZ) + P(XYZ) \tag{5}$$

In relation to events X and $\overline{X}$ *(not X)* the probability of their joint outcome will result

$$P(X + \overline{X}) = P(X) + P(\overline{X}) = 1 \tag{6}$$

It is possible to decompose a given event X introducing a given event Y and its complement $\overline{Y}$, so that

$$P(X) \equiv P(X \cap Y \cup \overline{Y}) = P(X \cap Y) + P(X \cap \overline{Y}) = P(X|Y)P(Y) + P(X|\overline{Y})P(\overline{Y}) \tag{7}$$

The above decomposition can be easily extended introducing more than one event. For instance,

$$P(X) = P(X|YZ)P(YZ) + P(X|Y\overline{Z})P(Y\overline{Z}) + P(X|\overline{Y}Z)P(\overline{Y}Z) + P(X|\overline{Y}\overline{Z})P(\overline{Y}\overline{Z}) \tag{8}$$

A. The Bayes Theorem

As is well known, the Bayes theorem answers the general problem on how to incorporate prior (in particular, subjective) knowledge into the analysis of an experiment and has been described by Bayes in a famous paper *An Assay Towards Solving a Problem in the Doctrine of Chances* published posthumously in 1763. It is derived by simple manipulation of joint, conditional, and marginal probabilities. If we denote by X_n the nth of N mutually exclusive events and by Y some other hypothesis or event, writing the probability of event X_n, given Y, as $P(X_n|Y)$ (as we have seen in previous section), according to Bayes theorem it is

$$P(X_nY) = P(X_n|Y)P(Y) = P(Y|X_n)P(X_n) \tag{9}$$

From this relation we obtain the equation

$$P(X_n|Y) = \frac{P(Y|X_n)P(X_n)}{\sum\limits_{m=1}^{N} P(Y|X_m)} \tag{10}$$

where we have replaced $P(Y)$ with $\sum\limits_{m=1}^{N} P(Y|X_m)$. This equation can be used for determining the probability $P(X_n|Y)$, once the probabilities appearing at the right hand side of Equation 10 are known, i.e., that corresponding to previous knowledge, $P(X_n)$, and those obtained from experimental data, $P(Y|X_m)$. The usefulness of this expression lies in the fact that in some cases it may be more convenient to do a sampling of event Y, given events X_m (m = 1,2, . . . N), rather than the other way around.

II. PROBABILISTIC ANALYSIS OF SYSTEM COMPONENTS

A system is a structure formed by one or more components, or units, or elements (generally subject to repair when failed). We shall limit here consideration to these components. The main concepts derived, however (as those relevant to the reliability and availability), can be directly extended to whole systems (or subsystems) as well.

We shall first distinguish between two types of components: those which operate on demand and those which operate continuously.

As far as the components operating on demand are concerned, we define the probability that at the l'th demand the component is in operation (event X_l). The probability that the component is in operatiuon at each of L demands can then be written as

$$P(X_1 X_2 \ldots X_L) = P(X_L | X_1 X_2 \ldots X_{L-1}) P(X_1 X_2 \ldots X_{L-1})$$

$$\equiv P(X_L | X_1 X_2 \ldots X_{L-1}) P(X_{L-1} | X_1 X_2 \ldots X_{L-2}) \ldots P(X_2 | X_1) P(X_1) \quad (11)$$

As far as the components operating continuously are concerned, let us indicate with F(t) the probability that the time of their first failure is less or equal than the time t of their continuous operation. Obviously, it is $0 \leqslant F(t) \leqslant 1$. The corresponding probability distribution function (pdf) f(t) will then be given by the equation

$$f(t) = \frac{dF(t)}{dt} \qquad (t > 0) \tag{12}$$

The quantity f(t)dt corresponds to the probability that the first failure occurs between t and t + dt.

We introduce now the mean time to failure (MTTF), i.e., the average time the component has functioned before failure. It will be given by the expression

$$MTTF = \int_0^\infty tf(t)dt \tag{13}$$

A. Component Reliability

The probability that in the interval (0,t) no failure (breakdown) occurs, in the sense that during this time the component keeps operating normally, is called "reliability" and is denoted with R(t). We easily obtain

$$R(t) = 1 - F(t) = \int_t^\infty f(t)dt \tag{14}$$

Clearly, $R(t) \rightarrow 0$ for $t \rightarrow \infty$. From the above expression, we see that the quantity F(t) corresponds to the complement of the reliability $[1 - R(t)]$. This quantity is called "unreliability" and is denoted $\bar{R}(t)$.

Since it is

$$\frac{dR(t)}{dt} = -f(t) \tag{15}$$

from Equation 13 we obtain that the MTTF is given by the expression

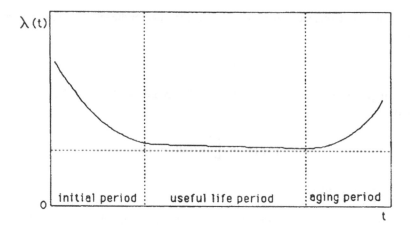

FIGURE 1.   Bathtub curve λ (t).

$$\text{MTTF} = \int_0^\infty R(t)dt \tag{16}$$

The (first) failure pdf f(t) for a component is normally expressed as the product

$$f(t) = \lambda(t)R(t) \tag{17}$$

where $\lambda(t)$ is the instantaneous failure rate, generally known as "hazard rate", and the quantity $\lambda(t)dt$ has the meaning of a conditional probability. In fact, it corresponds to the probability that such component is subject to failure at dt after having operated normally up to t.

Since it is $f(t) = -dR(t)/dt$, we obtain

$$\lambda(t) = -\frac{1}{R(t)}\frac{dR(t)}{dt} = \frac{f(t)}{R(t)} \tag{18}$$

and then the relationships

$$R(t) = e^{-\int_0^t \lambda(t)dt} \tag{19}$$

$$f(t) = \lambda(t)\, e^{-\int_0^t \lambda(t)dt} \tag{20}$$

Equation 19 is called the fundamental equation of reliability due to its importance in practical uses. In fact, the statistical data information on the reliability of components is generally given through values of $\lambda(t)$.

The time behavior of $\lambda(t)$ generally follows the so called "bath tub curve" (see Figure 1), i.e., characterized by three periods: a first period with $\lambda(t)$ on average high and then exponentially decreasing for the progressive elimination of all the weak units; a second period with $\lambda(t)$ relatively constant, called period of useful life, of larger or smaller extent; and a third period called aging period, during which $\lambda(t)$ increases rapidly following degradation processes. In practice, components are used only during the useful life period, i.e., after initial testings and before the expected start of the degradation process. Since in this case the λ value can be considered constant, the expression for the reliability can be simplified and results

$$R(t) = e^{-\lambda t} \tag{21}$$

Correspondingly, also the expression of the MTTF simplifies. Recalling Equation 15, it can be written

$$MTTF = \int_0^\infty R(t)dt = 1/\lambda \tag{22}$$

B. Component Availability

The reliability R(t) of a component gives the probability that it has not undergone failure up to time t and applies either to nonrepairable as well as repairable elements. We introduce now the concept of "availability" which applies to a component which, being repairable, can tolerate fault state conditions. In these cases the availability corresponds to the probability that such component is in operation at a given time t and is denoted as A(t). Its complement, $\overline{A}(t) = 1 - A(t)$ is called unavailability. Obviously it is $R(t) \leq A(t) \leq 1$, where the equation R(t) = A(t) applies for nonrepairable components.

Let us consider the case of a single component and introduce the instantaneous repair rate μ(t), whereas λ(t) corresponds to the instantaneous failure, or hazard, rate. If the instantaneous repair rate is constant, this amounts to assuming random repair times (not to be misinterpreted as times at which the repairing initiates). We have seen above that with constant λ the MTTF is given by 1/λ. In a quite analogous way, with constant μ, it can be found that the mean time to repair (MTTR, usually « MTTF) is given by 1/μ. The sum MTTR + MTTF corresponds to the mean time between failures (MTBF). Having assumed random instantaneous repair rates, with MTTR = 1/μ, it is easily found (e.g., using the maximum entropy law, see Section 2.E.3) that the repair time distribution is exponential, i.e., $\mu exp[-\mu(t - \bar{t}]$, $\bar{t}$ representing the time of failure from which the repairing can take place.

At asymptotic times, with constant values λ and μ, the availability will reach a constant value. This will be given by the ratio between the MTTF $(= \lambda^{-1})$ and the MTBF $(= \lambda^{-1} + \mu^{-1})$, i.e.,

$$A(\infty) = \frac{1/\lambda}{1/\lambda + 1/\mu} = \frac{\mu}{\lambda + \mu} \tag{23}$$

C. Failure Probability Distributions

We have seen above how the components forming a system may be distinguished between those operating on demand and those operating continuously. The first ones are subject to discrete fault events and are then described by discrete probability distributions, while the second ones are subject to continuous fault events and are then described by continuous probability distributions.

Among the discrete failure probability distributions the binomial and the Poisson distributions are particularly important. The binomial distribution is used for elements characterized by two possible states [event X/event $\overline{X}$ *(not* X), or up/down, or success/failure] in which the total number of trials (L) is known and in which their order does not influence the result. The Poisson distribution is used for elements in which random events produce irreversible transitions from one state to another.

1. Binomial Distribution

As mentioned above, the binomial distribution is of interest in case of components characterized by two possible states (success/failure). For instance, it can be of interest to determine the distribution of failures, out of a given total number of trials, relevant to the components operating on demand we have considered previously.

The binomial distribution can be obtained starting from the self-evident equation, relevant to the probability of the occurrence of event X (success) and of its complement $\overline{X}$ (failure), at each of L trials,

$$[P(X) + P(\overline{X})]^L = 1 \qquad (24)$$

i.e., corresponding to the (unit) probability expression relevant to all possible outcomes. The generic term at the left hand side

$$P(n) = \binom{L}{n} P(\overline{X})^n P(X)^{L-n} \equiv \frac{L!}{n!(L-n)!} P(\overline{X})^n P(X)^{L-n} \qquad (25)$$

represents the **probability** that in n trials (out of the total number of L) the outcome is a failure (event $\overline{X}$).

The mean value ($\bar{n}$) and the variance (σ^2) of n result

$$\bar{n} = \sum_{n=0}^{L} n\, P(n) = L\, P(\overline{X}) \qquad (26)$$

$$\sigma^2 = \sum_{n=0}^{L} (n - \bar{n})^2 P(n) = L\, P(\overline{X}) P(X) \qquad (27)$$

The cumulative probability distribution $P(n \leq x)$ that there occur a number of failures equal or less than x, out of a total number L of trials, can be easily obtained and is given by the relationship

$$P(n \leq x) = \sum_{n=0}^{x} P(n) \qquad (28)$$

The above derivation of the binomial distribution has been made considering binomial events (successes/failures) occurring at different trials on a single element. There is no difficulty in extending this distribution to events occurring simultaneously in a given number of identical components being tested at identical conditions.

2. Poisson Distribution

The Poisson distribution is obtained from the binomial one relevant to a very large number of identical trials for each of which the failure probability $P(\overline{X})$ is very small. For $L \to \infty$ and $P(\overline{X}) \to 0$ we easily obtain

$$P(n) = \frac{e^{-\mu}\, \mu^n}{n!} \qquad (29)$$

where here μ represents the average number of failures and results equal to its variance.

As we have seen with the binomial one, the Poisson distribution can be as well used to describe failure events occurring to a large number of identical components, which may be present in a given system at identical conditions, each with a very small failure probability $P(\overline{X})$.

3. Erlangian Distribution

The Erlangian distribution is a continuous distribution which is obtained from the Poisson one in which one sets the overall number of failures $\mu = \lambda t$, having assumed λ constant. It results

$$P(n,t) = \frac{e^{-\lambda t}}{n!} (\lambda t)^n \tag{30}$$

This distribution allows to determine the probability $f_n(t)dt$ that the nth component failure occurs in the interval dt at t, the element having been already subject to failure $(n - 1)$ times. It results

$$f_n(t)dt = \lambda P(n - 1, t)dt = \frac{\lambda (\lambda t)^{n-1} e^{-\lambda t} dt}{(n - 1)!} \tag{31}$$

If we set $n = 1$ in the above equation we have the (exponential) first failure distribution, i.e., setting f(t) in place of $f_1(t)$,

$$f(t) = \lambda e^{-\lambda t} \tag{32}$$

from which the average value $1/\lambda$ and the variance $1/\lambda^2$ of the time to failure can be obtained.

4. Gamma Distribution

The failure probability density as given by the gamma distribution is used to represent situations in which the component considered is subject to repetitive events (as thermal shocks, fatigue phenomena, etc.), so that the failure probability depends on the number of events produced. It is given by the expression

$$f(t) = \frac{\lambda (\lambda t)^{n-1} e^{-\lambda t}}{\Gamma(n)} \qquad (\lambda > 0; \quad n > 0) \tag{33}$$

where n may also be not an integer and where $\Gamma(n)$ is the gamma function. In case n is an integer, since in this case $\Gamma(n) \equiv n!$, we obtain again the Erlangian distribution.

5. Lognormal Distribution

Consider the normal gaussian distribution f(x) of a quantity x

$$y(x) = \frac{1}{\sqrt{2\pi}\,\sigma} e^{-(x-\bar{x})^2/2\sigma^2} \tag{34}$$

where $\bar{x}$ and σ^2 are the average value and the variance of x, respectively. If now ln t, its average value (expressed as lnβ) and its variance α^2 replace x, $\bar{x}$, and σ^2, respectively, we obtain the so-called lognormal distribution. It results

$$f(t) = \frac{1}{\sqrt{2\pi}\,\alpha t} \exp\left\{ - \frac{[\ln(t/\beta)]^2}{2\alpha^2} \right\} \tag{35}$$

where the presence of the 1/t coefficient at the right-hand side is due to the fact that the argument t, rather than ln t, is maintained, i.e., the distribution has been multiplied by

$\dfrac{d\ln t}{dt}$ ($= 1/t$). In this case the average value ($\bar{t}$) and the variance (σ^2) are given by the expressions

$$\bar{t} = \beta\, e^{-(\alpha^2/2)} \tag{36}$$

$$\sigma^2 = \beta^2\, e^{\alpha^2(e\alpha^2 - 1)} \tag{37}$$

The lognormal distribution is adequate in those cases in which rare events have to be distributed with rather poor information available.

6. Weibull Distibution

The Weibull distribution is used to describe any one of the three periods (initial, stationary, final) defined in relation to the bath tub curve of λ (see Figure 1), which we have encountered previously. It is given by the equation

$$f(t) = \lambda(t)\, e^{-\lambda(t)(t - \tau)/\beta} \qquad (t > \tau) \tag{38}$$

where $\lambda = \alpha\beta(t - \tau)^{\beta - 1}$, α is a scaling parameter and τ accounts for a delay time before failures initiate.

For $\beta < 1$ function $\lambda(t)$ exhibits a decreasing behavior. In this case it can be used to describe the initial period of a component. If $\beta = 1$, then $\lambda = \alpha$ and we obtain the exponential distribution. Finally, for $\beta = 2$, $\lambda(t)$ exhibits a rapidly growing behavior which can represent the third (final) period.

7. Other Distributions

We briefly describe in the following a few other failure probability distributions of some interest.

Mixed distributions — These distributions are formed by a combination of different probability distributions, i.e.,

$$f(t) = \sum_i k_i f_i(t) \tag{39}$$

where k_i and f_i are given coefficients and distribution functions, respectively.

Composite distributions — These distributions are synthetized by assigning different distributions in different time intervals into which the mission time has been subdivided.

Convoluted distributions — These, rather than to components, are relevant to standby devices (systems or subsystems). These devices often represent, however, irreducible units in system reliability studies (as in fault tree analysis) and for this reason they are mentioned here. Let us assume a standby device formed by I components. The device will operate until the last component is working. Perfect switching is assumed from one component to the next in case of failure. The probability density of i successive component failure at t, $f_{1,2,\ldots,i}(t)$, can be then given in terms of that relevant to the successive failures of the first $(i - 1)$ components, i.e.

$$f_{1,2,\ldots,i}(t) = \int_0^t f_i(t - t')\, f_{1,2,\ldots,(i-1)}(t')dt' \tag{40}$$

Function $f_{1,2,\ldots,(i-1)}$ can in turn be expressed by a similar convolution, and so on. For example, in the case of three components it is

$$f_{123}(t) = \int_0^t dt_2 f_3(t - t_2) \int_0^{t_2} dt_1 f_2(t_2 - t_1) f_1(t_1) \qquad (41)$$

8. Extreme Event Distributions

In order to define the probability distribution of extreme events, such as floods, droughts, disastrous earthquakes, etc., use is made of particular distributions obtained from considering those adopted for describing that same event at normal, i.e., not extreme, conditions and establishing a minimum, or maximum, value criterium.[2] For example, if in a river basin a flood is defined as the maximum daily discharge of the river in a solar year, from the statistical data available from observations of several years the flood size distribution over a large number of years can be evaluated.

D. Parameter Estimation

Given the experimental failure times data relevant to a sample of J identical components, we meet the problem of determining the failure time distribution for such a type of component. We shall describe in the following three methods widely used to this purpose: the moment method, the maximum likelihood method, and the maximum entropy method.

1. The Moment Method

Given the failure data obtained from a J component sample, the moment method consists in evaluating first the mean value and the variance of the failure time, using the expressions

$$\bar{t} = \frac{1}{J} \sum_{j=1}^{J} t_j \qquad (42)$$

$$\bar{\sigma}^2 = \frac{1}{N - 1} \sum_{j=1}^{J} (t_j - \bar{t}) \qquad (43)$$

where t_j represent the time of failure of the jth component. Once $\bar{t}$ and $\tilde{\sigma}^2$, together with the appropriate distribution curve (exponential, gamma, etc.), are given, the characterizing parameters (as we have seen from the component failure distributions, generally not exceeding the number of two) may then be estimated via the expressions which relate them to the average value and variance.

2. The Maximum Likelihood Method

The maximum likelihood method consists in determining those parameters which maximize the likelihood function

$$L(t_1, t_2, \ldots t_j | \theta_1, \theta_2, \ldots \theta_M) = \prod_{j=1}^{J} f(t_j | \theta_1, \theta_2, \ldots \theta_M) \qquad (44)$$

where $\theta_1, \theta_2, \ldots \theta_M$ represent the parameters to be estimated while $f(t_j | \theta_1, \theta_2, \ldots \theta_M)$ represents the chosen distribution function at the failure time t_j. Since the values $\tilde{\theta}_m$ which maximize L are the same which maximize lnL, in order to determine them it is preferable to start with this latter quantity. This amounts to set

$$\frac{\partial \ln L}{\partial \theta_m} = 0 \qquad (m = 1,2,\ldots M) \qquad (45)$$

The solution of this equation allows to obtain the estimates $\tilde{\theta}_n$. Their variance estimates result

$$\bar{\sigma}_m^2 = -\left(\frac{\partial^2 \ln L}{\partial \theta_m^2}\right)^{-1} \qquad (m = 1,2,...M) \qquad (46)$$

3. The Maximum Entropy Method

The idea of using information theory to predict the distribution of a set of random elementary events stems from the consideration that these, for the very fact of being random, tend to present themselves in the largest possible disorder (within given constraints), such that the knowledge of one of them would result most "informative".* Therefore, their distribution could be obtained maximizing this "informativeness", more precisely, maximizing the so-called information entropy function

$$S_I = -\frac{1}{\ln 2} \sum_i P_i \ln P_i \qquad (47)$$

where the sum is intended over all the possible outcomes and where P_i represents the probability of the outcome of the ith event X_i and should be intended in general as a conditional probability, i.e., $P_i = P(X_i|Y)$, where Y represents the given prior information (data estimates, constraints, etc.). Function S_I can be viewed as a measure of the missing information, corresponding to the number of questions (in terms of bits, i.e., yes/no, or true/false) foreseen to identify with certainty the true event, e.g., by subsequent halvings of the probability space. Given a number of possible events, it can be shown that S_I results maximum if they are equally distributed. Theoretical or experimental information which excludes some of the possibilities will obviously decrease it.

This law of maximum entropy was first formulated by Jaynes[4] in 1957. Its correct use presupposes that all (and only) the available theoretical and experimental information is to be exploited. It follows that the probability distributions obtainable according with this law depend on the choice of the (prior) data to be recorded. Its scope is quite general. In particular, it can be applied to determine the most likely distribution of a set of possible values associated with a given parameter, either basing on measurements, or on personal, coherent judgment.

Consider now the case of J identical components (this implying also a common failure distribution) subject to testing. The failure times result t_j $(j = 1,2, \ldots J)$, from which sample estimates, e.g., of the mean and variance, can be made.

If we subdivide the time in equal intervals $\delta t_r = (t_{r+1} - t_r)$, having set $t_o = 0$, and we define P_r as the probability that a component under test fails in the interval δt_r, then Equation 47 can be written, apart from a constant coefficient,

$$S_I = -\sum_{r=1}^{\infty} P_r \ln P_r \qquad (48)$$

Assuming the obvious constraint

$$\sum_{r=1}^{\infty} P_r = 1 \qquad (49)$$

and others of the general type

$$\sum_{r=1}^{\infty} T_m(t_r) P_r = \bar{T}_m \qquad (m = 1,2,...M) \qquad (50)$$

---

* If we considered the opposite, limiting case of a set of nonrandom, i.e., predetermined, events, the information contained in the detection (in this case, verification) of one of them would appear obviously null, their occurrence being known in advance.

T_m and $\bar{T}_m$ representing M assigned functions and their average values (of which estimates are known), respectively, the function, F, to be maximized will result

$$F = - \sum_{r=1}^{\infty} P_r \ln P_r + (k_0 - 1)\left(\sum_{r=1}^{\infty} P_r - 1\right) +$$

$$k_1\left(\sum_{r=1}^{\infty} T_1(t_r)P_r - \bar{T}_1\right) + \ldots k_M\left(\sum_{r=1}^{\infty} T_M(t_r)P_r - \bar{T}_M\right) \qquad (51)$$

where $(k_0 - 1)$, k_1, ... k_M are the Lagrange multipliers. By differentiating with respect to P_r, the following equations are then obtained

$$\frac{dF}{dP_r} = -\ln P_r + k_0 + k_1 T_1(t_r) + \ldots + k_M T_M(t_r) = 0 \qquad (r = 1,2,\ldots) \qquad (52)$$

from which

$$P_r = e^{-k_0 - k_1 T_1(t_r) - \ldots - k_M T_M(t_r)} \qquad (r = 1,2,\ldots) \qquad (53)$$

where the values of the Lagrange multipliers k_0, k_1, ... k_M are obtained using the constraints, Equations 49 and 50.

If in place of discrete probability values P_r we set $\bar{f}_r \delta t_r$, where $\bar{f}_r$ is the average probability density of component failure within interval δt_r, and then we make $\delta t_r \to 0$, we obtain the (continuous) probability density

$$f(t_r) = \lim_{\delta t_r \to 0} \bar{f}_r$$

$$= e^{-k_0 - k_1 T_1(t_r) - \ldots - k_M T_M(t_r)} \qquad (t_r \geq 0) \qquad (54)$$

As said above, the values k_0, k_1, ... k_M can be obtained using the constraints, which in this case will correspond to Equations 49 and 50 with the sum replaced by the corresponding integral. We can then write

$$\int_0^{\infty} f(t)dt = e^{-k_0} \int_0^{\infty} e^{-k_1 T_1(t) - \ldots - k_M T_M(t)} \, dt = 1 \qquad (55)$$

$$\int_0^{\infty} T_m(t)f(t)dt = e^{-k_0} \int_0^{\infty} T_m(t) \, e^{-k_1 T_1(t) - \ldots - k_M T_M(t)} \, dt$$

$$= \bar{T}_m \qquad (m = 1,2,\ldots M) \qquad (56)$$

From these equations we easily have

$$\frac{dk_0}{dk_m} = \frac{dk_0}{d[\exp(-k_0)]} \frac{d[\exp(-k_0)]}{dk_m} = -\int_0^{\infty} T_m f(t)dt = -\bar{T}_m \qquad (m = 1,2,\ldots M) \qquad (57)$$

Now, there are two possibilities, depending on whether the specific distribution function relevant to the events considered is unknown, or the type of it is given and we wish to determine the parameters which define it. In the first case, the information represented by estimates of the quantities $\bar{T}_m$ (generally the moments t^a) will allow to determine the values

k_0, k_1, . . . k_M. In the second case, it will be necessary to express the given function in the form, Equation 54, and then, using Equation 57, determine its parameters.

For illustration, let us consider a quantity of which we know only an estimate of its mean value. Assume that this quantity is the MTTF $(\bar{t})$ of a component. In this case $M = 1$ and we can use only two constraints. The corresponding Lagrange multipliers can be obtained from the two equations

$$e^{-k_0} \int_0^\infty e^{-k_1 t} \, dt = \frac{e^{-k_0}}{k_1} = 1 \tag{58}$$

$$e^{-k_0} \int_0^\infty t e^{-k_1 t} \, dt = \frac{e^{-k_0}}{k_1^2} = \frac{1}{k_1} = \bar{t} \tag{59}$$

It then results $k_1 = 1/\bar{t}$ and $k_0 = -\ln(1/\bar{t})$. The distribution function then will be, setting λ in place of $1/\bar{t}$,

$$f(t) = \lambda e^{-\lambda t} \tag{60}$$

i.e., we find that the exponential distribution is the most probable one in absence of supplementary information.

It is also important to verify that, had we chosen a priori the exponential distribution as that relevant to the events considered, we would have found that $\lambda = 1/\bar{t}$.

If, besides the estimate of the mean value $\bar{t}$, we have had also an estimate of the variance but no information on higher order moments nor on the type of curve to be expected, we would have obtained the well-known normal Gaussian distribution. If, instead, beside the first and second moments, we have had reasons to believe that the exponential function were more appropriate, based on personal judgment founded on experience on previous similar cases, the information on the variance would have to be dropped. In fact, with the exponential distribution, as we have seen in previous sections, the variance is given by the square of the mean value (in this case $\bar{t}^2$). In this case, the information on the variance could be rather used to verify the validity of the exponential distribution choice.

III. REPRESENTATION OF SYSTEMS

As mentioned previously, a system is a structure consisting of a number of basic interconnected components functioning in such a way that a specific mission can be accomplished, given initial conditions and constraints.

Systems can be subject to various classifications in relation to the criteria chosen. In particular, they can be classified as statistically coherent (s-coherent) or noncoherent. A system is called s-coherent if, in whatever state it may be, the good performance of one of its components always contributes to the correct functioning of the whole system. In noncoherent systems there are states in which this condition may not be so, i.e., the correct functioning (e.g., after a repair) of a component may make the system fail.

S-coherent systems have been prevalently studied until a few years ago, but in later years there has been an increasing interest also for noncoherent ones.[6-8] These can in fact be encountered in practice. A significant example is represented by nuclear safety systems, due to the regulatory requirement that not all the redundant loops should be simultaneously disabled due to maintenance.[9]

In order to simplify the treatment, in the following we shall always refer to s-coherent systems, when not differently specified. Extension of the formulation to noncoherent ones will then be treated in a separate section.

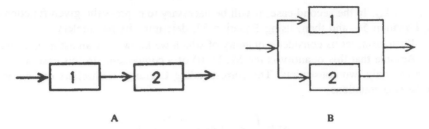

FIGURE 2.   (A) System with two elements in series and (B) system with two elements in parallel.

A. Block Diagram Representation

Systems can be represented by block diagrams. A block diagram consists of a set of blocks (squares) representing the various components of the system considered, linked with each other according to their various interconnections. For illustration, in Figures 2A and B two block diagrams are shown relevant to two simple systems formed each by two components operating in series and in parallel, respectively.

If we interpret these blocks as switches which can be closed (component working), or open (component failed), and the diagram as an electric circuit with input and output we can say that the system is operating (i.e., in up state) if there is at least one continuous path connecting the input and the output through which the current can run. Otherwise, the system would result failed.

The electric circuit interpretation of the block diagram has been motivated by its historical origin, since it was first used for reliability studies on the electric circuits. Presently there is the tendency to interpret the block diagram as a logical representation. To each block, representative of a given component, one of two possible states (events) are associated: state 1, or up (component working) and state 0, or down (component failed). Then the procedures of the Boolean algebra (called also algebra of events) are applied (see Appendix A). By these procedures the binary operations of intersection ($\cap$) and union ($\cup$) between two events associated with two blocks correspond to the event resulting from their connection in series (conjunction: and) and parallel (disjunction: or), respectively.*

Based on the above interpretation, we can then say that a system represented by a block diagram will be operating (up) if in such scheme there exists at least one continuous path between input and output, such that the Boolean intersection of all the events corresponding to the blocks encountered will result equal to unity. The Boolean intersections satisfying this condition of continuity between input and output of a block diagram are called path sets and will be generically denoted with the symbol G_r^+ ($r = 1,2, \ldots$).

Let us denote with X_n ($n = 1,2, \ldots N$) the Boolean indicator relevant to the generic nth component, N being the total number of components in the system considered. We can then write

$$G_r^+ = \bigcap_{X_n \in G_r^+} X_n \qquad (61)$$

When a path set is such that the failure of any one of the corresponding components

---

*   In closer analogy with the electric circuit interpretation, sometimes, although not very frequently, rather than block diagrams, the so-called signal flow graphs are used. These graphs consist of branches, through which the signal flows from left to right, and nodes, at which branches join. Each branch corresponds to a component and the signal can flow through it only if the component is at up state. The system is at up state if there is a path in the graph through which the signal can flow from left to right. It is evident that the failure of a set of components, such that the graph results split into two noncommunicating subgraphs, will make the system fail. For example, in Figures 3A and B the signal flow graphs are shown corresponding to the two block diagrams of Figures 2A and B (relevant to two systems with two elements in series and in parallel, respectively.)

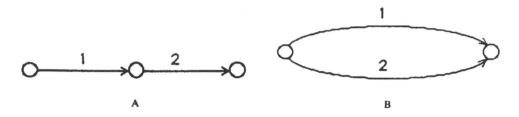

FIGURE 3. (A) Signal flow graph for a system with two elements in series and (B) signal flow graph for a system with two elements in parallel.

interrupts its continuity between input and output of the block diagram, independently from the state of the other components, i.e., when its Boolean expression in such case is consistently zero, then it is called a minimal path set. A minimal path set will be denoted G_l ($l = 1, 2, \ldots L$), L being their total number.

Any path set can in general be reduced, by application of Boolean algebra, to a minimal path set. If by $x_{l,i}$ ($i = 1, 2, \ldots I_l$), where I_l is the number of components comprised in the lth minimal path set, we denote the Boolean indicators of such components, then G_l result given by their intersection. So, it can be written

$$G_l = X_{l,1} \cap X_{l,2} \cap \ldots X_{l,I_l} = \bigcap_{i=1}^{I_l} X_{l,i} \qquad (l = 1, 2, \ldots L) \tag{62}$$

If we now introduce vector X to represent the state (of up or down condition) or all the components and, consequently, the up or down condition of the system, we can generally define a Boolean variable $\Phi(X)$, denoted as structure function, indicative of the system state, i.e., with values 1 or 0, depending on whether the system is operating (up), or failed (down), respectively. The following irreducible expression can then be written

$$\Phi(X) = G_1 \cup G_2 \cup \ldots G_L \equiv \bigcup_{l=1}^{L} G_l \tag{63}$$

where the path sets G_s are given by Equation 62.

For simplicity, from now on in explicit Boolean expressions we shall make use of the usual multiplication ($\cdot$, normally omitted) and sum ($+$) notations in place of those of intersection ($\cap$) and union ($\cup$). With this new notation the above expression for ϕ, Equation 63, recalling Equation 62, can then be written

$$\Phi(X) = G_1 + G_2 + \ldots + G_L = \bigcup_{l=1}^{L} X_{l,1} X_{l,2} \ldots X_{l,I_l} \tag{64}$$

Using Boolean algebra properties, we can easily determine the complement of ϕ, which we shall denote as $\overline{\Phi}$, i.e.,

$$\overline{\Phi}(X) = \overline{G}_1 \overline{G}_2 \ldots \overline{G}_L = \bigcap_{l=1}^{L} (\overline{X}_{l,1} + \overline{X}_{l,2} + \ldots \overline{X}_{l,I_l}) \tag{65}$$

We have seen above how the continuity of a block diagram from input to output can be described by the path sets. We shall introduce now the concept of cut set. This is defined as the Boolean intersection of a given series of complemented events corresponding to a set

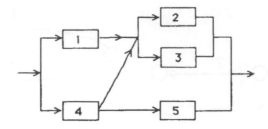

FIGURE 4.   Block diagram example.

of components such that, if they all fail, the system fails. We shall generally denote them by the symbol $\overline{C}_t^+$ (t = 1,2, . . .). We can then write

$$\overline{C}_t^+ = \bigcap_{X_n \in \overline{C}_t^+} X_n \tag{66}$$

When a cut set is such that the normal functioning of any one of the corresponding components guarantees the normal functioning of the system, independently from the state of the other components, i.e., when its Boolean expression in such case is consistently zero, then it is called a minimal cut set. A minimal cut set will be denoted $\overline{C}_k$ (k = 1,2, . . . K), K being their total number.

Any cut set can in general be reduced, by application of Boolean algebra, to a minimal cut set. If by $X_{k,j}$ (j = 1,2, . . . J_k), where J_k is the number of components comprised in the kth minimal cut set, we denote the Boolean indicators of such components, then, the cut set $\overline{C}_k$ results given by the intersection of their complements. So, it can be written

$$\overline{C}_k = X_{k,1}X_{k,2} \ldots X_{k,J_k} = \bigcap_{j=1}^{J_k} X_{k,j} \quad (k = 1,2,\ldots K) \tag{67}$$

The Boolean complement of the structure function can now be expressed in terms of the minimal cut sets by the following irreducible expression

$$\overline{\Phi}(X) = \overline{C}_1 + \overline{C}_2 + \ldots + \overline{C}_K = \bigcup_{k=1}^{K} X_{k,1}X_{k,2}\ldots X_{k,J_k} \tag{68}$$

which is equivalent to Equation 65.

Let us consider the simple example of a system formed by five components, as illustrated in Figure 4. Its structure function will result

$$\Phi(X) = X_1X_2 + X_1X_3 + X_4X_2 + X_4X_3 + X_4X_5 \tag{69}$$

and its complement

$$\overline{\Phi}(X) = (\overline{X}_1 + \overline{X}_2)(\overline{X}_1 + \overline{X}_3)(\overline{X}_4 + \overline{X}_2)(\overline{X}_4 + \overline{X}_3)(\overline{X}_4 + \overline{X}_5)$$
$$= \overline{X}_1\overline{X}_4 + \overline{X}_2\overline{X}_3\overline{X}_4 + \overline{X}_2\overline{X}_3\overline{X}_5 \tag{70}$$

where X_1X_2, X_1X_3, X_4X_2, X_4X_3, and X_4X_5 correspond to the minimal path sets G_1, G_2, G_3, G_4, and G_5, while $\overline{X}_1\overline{X}_4$, $\overline{X}_2\overline{X}_3\overline{X}_4$, and $\overline{X}_2\overline{X}_3\overline{X}_5$ correspond to the minimal cut sets $\overline{C}_1$, $\overline{C}_2$, and $\overline{C}_3$, respectively.

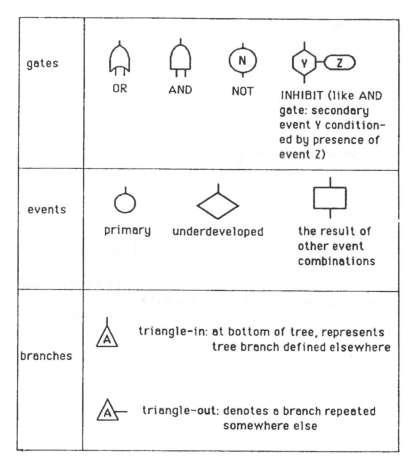

FIGURE 5. Fault tree symbols.

B. Fault Tree Representation

The complement $\bar{\Phi}(X)$ of the structure function relevant to a given system can be represented by a diagram called "fault tree", in which a set of special symbols is adopted (see Figure 5), among which those representing intersection and union of events and denoted "and gate" and "or gate", respectively. In the fault tree terminology the event corresponding to $\bar{\Phi} = 1$ (system at failed state, or down) is denoted as "top event".*

To familiarize with the basic symbology used in the representation of events, in Figures 6A and B the correspondence between the block diagram and the fault (and success) tree representations is shown, relevant to two simple systems with two elements in series and in parallel, respectively.

The construction of a fault tree can be easily done, once the minimal cut sets are known. For illustration, the fault tree relevant to the example considered in previous section (Figure 4) is shown in Figure 7A. In a completely resolved fault tree the branching ends at bottom with primary events, corresponding to the system components. In other cases, the branching may exhibit at its bottom undeveloped events (e.g., subsystems).

Fault tree levels are defined, depending on the number of branchings between the top event and the fault event (of components, or subsystems) considered. In the example above

---

* As we shall see in the following, to a fault tree it is always possible to associate the corresponding success tree representing the structure function ϕ and so called since in this case the top event ($\phi = 1$) corresponds to system in upstate.

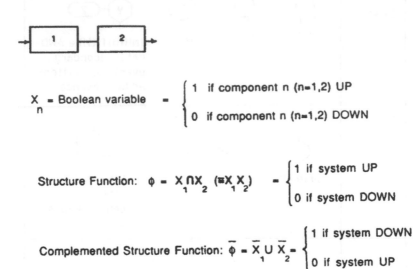

Block Diagram

X_n = Boolean variable = $\begin{cases} 1 & \text{if component n (n=1,2) UP} \\ 0 & \text{if component n (n=1,2) DOWN} \end{cases}$

Structure Function: $\phi = X_1 \cap X_2$ ($=X_1 X_2$) = $\begin{cases} 1 & \text{if system UP} \\ 0 & \text{if system DOWN} \end{cases}$

Complemented Structure Function: $\bar{\phi} = \bar{X}_1 \cup \bar{X}_2$ = $\begin{cases} 1 & \text{if system DOWN} \\ 0 & \text{if system UP} \end{cases}$

Success Tree   Fault Tree

A

FIGURE 6.   (A) Representations of a system with two elements in series. (B) Representations of a system with two elements in parallel.

(Figure 7A), for instance, we can distinguish: level 0, in which we find the top event $\bar{\phi}(X)$; level 1, in which we find events $(\bar{X}_1 \cap \bar{X}_4)$ and $[\bar{X}_2 \cap \bar{X}_3 \cap (\bar{X}_4 \cup \bar{X}_5)]$; level 2, in which we find event $[\bar{X}_3 \cap (\bar{X}_4 \cup \bar{X}_5)]$ and primary events $\bar{X}_1, \bar{X}_2, \bar{X}_4$; level 3, in which we find event $(\bar{X}_4 \cup \bar{X}_5)$ and primary event $\bar{X}_3$; and level 4, in which we find primary events $\bar{X}_4$ and $\bar{X}_5$.

Sometimes different, qualitative, fault tree level classifications are defined, as top structure level, major system levels, subsystem level, and detailed component level.

Given a fault tree, it is possible to obtain from it the (dual) "success tree" substituting the and gates with the or gates, and vice versa, and complementing the events. The top event will then represent the structure function $\phi(X)$, of value 1 in case of success and 0 otherwise. The construction of a success tree can also be made directly, once the minimal path sets are known.

For illustration, the success tree relevant to the example above, i.e., dual to fault tree,

Block Diagram

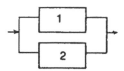

X_n - Boolean variable $=$ $\begin{cases} 1 & \text{if component n (n=1,2) UP} \\ 0 & \text{if component n (n=1,2) DOWN} \end{cases}$

Structure Function: $\phi = X_1 \cup X_2 \ (=X_1 + X_2)$ $=$ $\begin{cases} 1 & \text{if system UP} \\ 0 & \text{if system DOWN} \end{cases}$

Complemented Structure Function: $\bar{\phi} = \bar{X}_1 \cap \bar{X}_2 =$ $\begin{cases} 1 & \text{if system DOWN} \\ 0 & \text{if system UP} \end{cases}$

Success Tree   Fault Tree

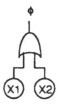   

FIGURE 6B.

Figure 7A, is shown in Figure 7B. The primary events in this case are represented by the uncomplemented indicators X_1, X_2, X_3, X_4, and X_5. Making use of the path sets evidenced by it, the corresponding block diagram can be inversely constructed, as shown in Figure 8. It is easy to verify, for example, by writing the Boolean expression of the structure function $\phi(X)$, that it is exactly equivalent, as it should be, to the original one, Figure 4.

In the above fault tree description, its relationship with the block diagram representation, from which it can be derived, was evidenced.* In practical cases, however, which are relevant to complex systems with a very high number of components, the fault tree is constructed inductively, based on a detailed knowledge of the system design and of its safety devices and criteria. The main reason for its construction is in fact the identification of the minimal cut sets themselves. These can be generally determined expressing the comple-

* If the block diagram of a system is reduced in the form of the type illustrated in Figure 8, i.e., with only connections in series and parallel between blocks (including already connected ones), it is instructive to note that, to construct the corresponding fault tree, we can follow this simple procedure: start with such block diagram, then (recalling the correspondence of the parallel and series connections with the or and and gates, respectively) construct the success tree and finally, by the usual event and gate complementation, the fault one.

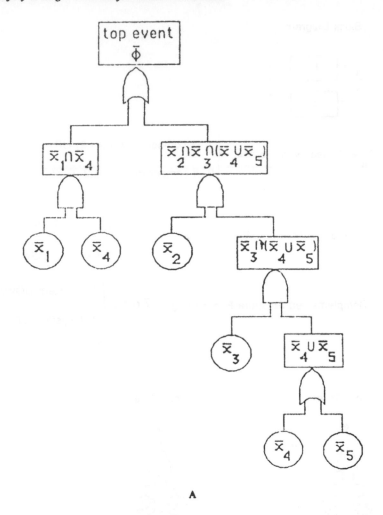

A

FIGURE 7.   (A) Fault tree example and (B) success tree example.

mented structure function (top event) of the system in terms of the primary fault events and then reducing the expression so obtained making use of the Boolean algebra properties. For complex systems the search of the minimal cut sets along with this procedure is made with the use of a computer.[12-14]

To further illustrate the fault tree representation, another example is shown in Appendix B.

IV. PROBABILISTIC ANALYSIS OF SYSTEMS

To the generic Boolean variable X_n (n = 1,2, . . . N) relevant to a given component, its expected value $E\{X_n\}$ as a function of time can be associated, or, which is the same, the probability $P(X_n = 1|t)$ [or $P(X_n|t)$]** at t (corresponding to the component availability defined in Section 2). Through the Boolean relationship relating the system structure function $\phi(X)$ with the component Boolean indicative variables X_n, the expected value of $\phi(X)$ itself, $E\{\phi(X)\}$ [denoted also as probability $P(\phi)$], or that, $E\{\overline{\phi}(X)\}$ [$\equiv P(\overline{\phi})$], of its complement

** From now on, notation $P(X = 1)$ [$\equiv E\{X\}$], X being a Boolean variable, will be simplified into $P(X)$.

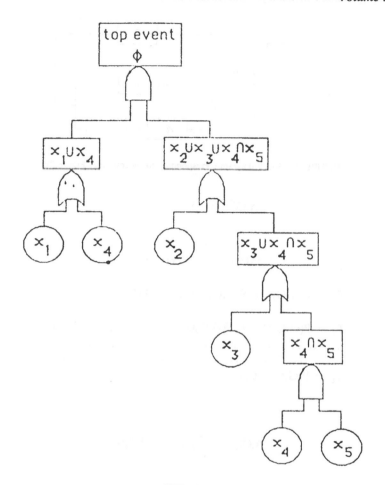

FIGURE 7B.

$\overline{\Phi}(\mathbf{X})$, can be obtained using the above expressions of $\phi(\mathbf{X})$, or $\overline{\phi}(\mathbf{X})$, together with the expressions of joint, or disjoint, probabilities described in Section I.

A. System Availability and Unavailability

Probabilities $P(\phi)$ and $P(\overline{\phi})$ relevant to the system structure function and its complement define the system availability [which we shall denote with symbol $A_s(\mathbf{X})$] and unavailability [which we shall denote with symbol $\overline{A}_s(\mathbf{X})$], respectively. We can then write

$$A_s(\mathbf{X}|t) = E[\Phi(\mathbf{X}|t)] \tag{71}$$

$$\overline{A}_s(\mathbf{X}|t) = E[\overline{\Phi}(\mathbf{X}|t)] \equiv 1 - A_s(\mathbf{X}|t) \tag{72}$$

where we have also indicated the general dependence on t ($0 \leqslant t \leqslant t_F$, t_F being the mission time).

Recalling the expressions of joint and disjoint probabilities described in Section I, and Equations 64 and 68, giving ϕ and $\overline{\phi}$ as function of $\mathbf{X}_n$, or their complements, respectively, the expressions of A_s and $\overline{A}_s$ in terms of the probability of occurrence of path sets, or cut sets, can be obtained. Recalling Equations 64, 68, 71, and 72, it results

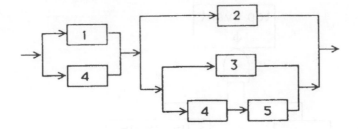

FIGURE 8.   Block diagram corresponding to success tree, Figure 7B.

$$A_s(\mathbf{X}|t) = \sum_{l=1}^{L} (-1)^{l-1} T_l \tag{73}$$

with

$$T_1 = P(G_1) + P(G_2) + ... + P(G_L) \tag{74}$$

$$T_2 = P(G_1G_2) + P(G_1G_3) + ... + P(G_{L-1}G_L) \tag{75}$$

$$T_3 = P(G_1G_2G_3) + P(G_1G_2G_4) + ... + P(G_{L-2}G_{L-1}G_L) \tag{76}$$
$$\dotfill$$
$$\dotfill$$
$$T_L = P(G_1G_2...G_L) \tag{77}$$

and

$$\overline{A}_s(\mathbf{X}|t) = \sum_{k=1}^{K} (-1)^{k-1} \hat{T}_k(t) \tag{78}$$

with

$$\hat{T}_1 = P(\overline{C}_1) + P(\overline{C}_2) + ... + P(\overline{C}_K) \tag{79}$$

$$\hat{T}_2 = P(\overline{C}_1\overline{C}_2) + P(\overline{C}_1\overline{C}_3) + .. + P(\overline{C}_{K-1}\overline{C}_K) \tag{80}$$

$$\hat{T}_3 = P(\overline{C}_1\overline{C}_2\overline{C}_3) + P(\overline{C}_1\overline{C}_2\overline{C}_4) + ... + P(\overline{C}_{K-2}\overline{C}_{K-1}\overline{C}_K) \tag{81}$$
$$\dotfill$$
$$\dotfill$$
$$\hat{T}_K = P(\overline{C}_1\overline{C}_2...\overline{C}_K) \tag{82}$$

with G_l and $\overline{C}_k$ given by Equations 62 and 67, respectively.

The following obvious relationships can also be written, recalling expressions (Equations 73 and 78)

$$\overline{A}_s(\mathbf{X}|t) = 1 - A_s(\mathbf{X}|t) = 1 - \sum_{l=1}^{L} (-1)^{l-1} T_l(t) \tag{83}$$

$$A_s(\mathbf{X}|t) = 1 - \overline{A}_s(\mathbf{X}|t) = 1 - \sum_{k=1}^{K} (-1)^{k-1} \hat{T}_k(t) \tag{84}$$

B. System Reliability

As we have seen in previous section, the system availability $A_s(t)$ and the system unavailability $\overline{A}_s(t)$ at time t can be evaluated once the minimal cut sets and the associated probabilities at t are known. This is not generally so with the system reliability and system unreliability, for the evaluation of which the knowledge of the entire system history up to t is needed. For complex systems a precise evaluation of the system reliability, which we shall denote as $R_s(t)$, can be generally obtained only if we can solve the Markov chain problem associated with the evolution of the system considered (see Section V). Upper and lower bounds, however, can be obtained, as shown in Section D.

1. Examples

For illustration, we shall consider a few examples of simple systems.

Consider first a system of N components connected in series. Since in this case the structure function is clearly $\phi(X) = \prod_{n=1}^{N} X_n$, the system reliability will be, the events X_n having been assumed independent from each other,

$$R_s(t) = \prod_{n=1}^{N} R_n(t) \tag{85}$$

$R_n(t)$ representing the nth component reliability. Correspondingly, the MTTF will result

$$MTTF = \int_0^\infty R_s(t)dt = \left(\sum_{n=1}^{N} \lambda_n \right)^{-1} \tag{86}$$

which for identical units reduces to $MTTF = (N\lambda)^{-1}$

Since $R_s < 1$, clearly the reliability decreases with increasing number of components (tending to 0 with n → ∞) and, correspondingly, the MTTF decreases.

Consider then a system of N nonrepairable components connected in parallel. Since in this case the structure function is clearly $\phi(t) = 1 - \overline{\phi}(t) = 1 - \prod_{n=1}^{N} X_n$, the system reliability will be, the events X_n having been again assumed independent from each other,

$$R_s(t) = 1 - \overline{R}_s(t) = 1 - \prod_{n=1}^{N} \overline{R}_n(t) \tag{87}$$

Clearly, $R_s(t)$ increases with N (tending to 1 for N → ∞). For identical units the mean time to failure (MTTF) results

$$MTTF = \sum_{n=1}^{N} (n\lambda)^{-1} \tag{88}$$

In the case of a system formed by N identical, not repairable, components connected in parallel, and working with at least M (<N) of them in operation, the corresponding structure function will be given by the union of the Boolean intersections corresponding to all the combinations of M events out of N, i.e.,

$$\Phi(X) = \bigcup_{i_1 < i_2 < \ldots < i_M = 1}^{N} X_{i_1} X_{i_2} \ldots X_{i_M} \tag{89}$$

The corresponding reliability results

$$R_s(t) = \sum_{n=M}^{N} \frac{N!}{n!(N-n!)} [R(t)]^n [1 - R(t)]^{N-n} \tag{90}$$

where $R(t)$ is the component reliability.

Let us now consider again the example described in Section III.A by the block diagram, Figure 4. In this case the structure function is given by Equation 69. Assuming again that the components are independent and not repairable and recalling Equation 73, the following expression for the system reliability can be obtained

$$
\begin{aligned}
R_s(t) &= E\{X_1X_2 + X_1X_3 + X_2X_4 + X_3X_4 + X_4X_5\} \\
&= R_1R_2 + R_1R_3 + R_2R_4 + R_3R_4 + R_4R_5 - \\
&\quad - R_1R_2R_3 - R_1R_2R_4 - R_1R_3R_4 - R_2R_3R_4 - \\
&\quad - R_2R_4R_5 - R_3R_4R_5 + R_1R_2R_3R_4 + R_2R_3R_4R_5
\end{aligned}
\tag{91}
$$

The above expression can be reduced into a more manageable form if the so-called decomposition method is used, as illustrated in the following.

C. Decomposition Method

The decomposition method corresponds to identifying a "key element" in the block diagram of a given system such that either its shortening (i.e., connecting its input and output) or its opening (i.e., cutting the diagram in correspondence of it) result in a block diagram simpler to analyze. Once a key element has been identified, e.g., the kth one, the decomposition expression is then obtained, corresponding to Equation 7 with X representing the structure function (ϕ) and Y the event (X_k) corresponding to the key element.

If, for illustration, in the example considered in previous section relevant to the block diagram of Figure 4, we chose as key component that corresponding to block 4, we recognize that its shortening and opening results in two simpler block diagrams corresponding, respectively, to a three-parallel component system (elements 2, 3, and 5) and to a system in which element 1 is in series with elements 2 and 3 in parallel. In place of Equation 91 we can then write the equivalent expression

$$R_s(t) = R_4[1 - (1 - R_2)(1 - R_3)(1 - R_5)] + (1 - R_4)[R_1(R_2 + R_3 - R_2R_3)] \tag{92}$$

D. Bounding Methods

As we mentioned in Section IV.B, for the evaluation of the reliability R_s relevant to complex systems, lower and upper bounds are usually estimated.

An upper bound for $R_s(t)$ is of course the availability $A_s(t)$ (the system may have failed before time t, even though at t is operating), whereas a lower bound is obtained evaluating the availability of the same system assuming all its components with infinite repair time (i.e., not repairable). If we denote this latter availability (reliability) as $R_s^{NR}(t)$, we can then write

$$R_s^{NR}(t) \leq R_s(t) \leq A_s(t) \tag{93}$$

$R_s^{NR}(t)$ can be determined similarly as with $A_s(t)$, where the component availabilities $A_n(t)$ ($n = 1,2, \ldots N$) are replaced by the corresponding reliabilities $R_n(t)$.

Lower and upper bounds can be of interest also with respect to the availability (or unavailability). In fact, for complex systems with a large number of components, calculating all terms T_i, or $\hat{T}_i$, in the sums of Equations 73, or 78, respectively, may become unmanageable. In the following, a few commonly used bounding methods are shown.

1. Inclusion-Exclusion Method

It is obtained from Equations 73 and 78 arresting the sums over l and k to an assigned even number of terms T_l and $\hat{T}_k$ [e.g., L' ($\leq$L) and K' ($\leq$K), respectively], this giving a lower bound, and successively arresting the above sums to the same number of terms plus one, this giving an upper bound.

Rather than from Equations 73 and 78, we can obtain bounding values from Equations 84 and 83, respectively. The bounding procedure is quite analogous to that described above, with the advertence that L' and K' should be now chosen uneven.

A particular application of the inclusion-exclusion method is that in which use is made of the unavailability expression, Equation 78, where the sum is limited to the first term (i.e., K' = 1). It is obtained

$$0 < \overline{A}_s(t) \leqslant \hat{T}_1(t) \tag{94}$$

This expression corresponds to the so-called "rare event approximation," since it is generally considered to give acceptable estimates of the system unavailability when the unavailabilities of the components, $\overline{A}_n(t)$, are very small (e.g., not exceeding the value of 0.1).

2. Min-Max Method

Another bounding expression can be obtained for the system unavailability $\overline{A}_s(t)$ if we consider expressions of the complemented function $\overline{\Phi}$ in terms of cut sets and complemented minimal path sets, respectively.

Let us first recall Equation 68, expressing the complemented structure function $\overline{\Phi}$ in terms of cut sets, i.e.,

$$\overline{\Phi}(\mathbf{X}) = \bigcup_{k=1}^{K} \bigcap_{j=1}^{J_k} X_{k,j} \tag{95}$$

We can easily verify the following relationship

$$\overline{A}_s = P(\overline{\Phi}) = P\left(\bigcup_{k=1}^{K} \bigcap_{j=1}^{J_k} X_{k,j} \right) \geqslant \max\left[\prod_{j=1}^{J_k} P(X_{k,j}) \right] \tag{96}$$

Let us then recall Equation 65, expressing now the complemented structure function in terms of the complemented path sets, i.e.,

$$\overline{\Phi}(\mathbf{X}) = \bigcap_{l=1}^{L} \bigcup_{i=1}^{l_1} X_{l,i} \tag{97}$$

We can then easily verify that the following relationship holds

$$\overline{A}_s = E\{\overline{\Phi}\} = P\left(\bigcap_{l=1}^{L} \bigcup_{i=1}^{l_1} X_{l,i} \right) \leqslant \min\left[\sum_{i=1}^{l_1} P(X_{l,i}) \right] \tag{98}$$

The following min-max bounding expression can therefore be written

$$\text{Max}\left[\prod_{j=1}^{J_k} P(X_{k,j}) \right] \leqslant \overline{A}_s \leqslant \min\left[\sum_{i=1}^{l_1} P(X_{l,i}) \right] \tag{99}$$

3. Min-Path and Min-Cut Bound Method

This method again allows to calculate lower and upper bounds relevant to the system unavailability $\overline{A}_s$. It is obtained considering, as above, Equations 65 and 68. Recalling again Equations 95 and 97, giving the complemented structure function in terms of the cut sets and complemented path sets, respectively, the following expression is obtained

$$\prod_{l=1}^{L} P\left(\bigcup_{i=1}^{I_l} X_{l,i}\right) \le \overline{A}_s \le \sum_{k=1}^{K} P\left(\bigcap_{j=1}^{J_k} X_{k,j}\right) \tag{100}$$

The right upper bound is evident (no compensation terms being considered relevant to the probability of the union over the cut sets). The left lower bound is obtained considering that the product of the probabilities of the unions of events $X_{l,i}$ is smaller than the probability of the corresponding intersection. This is so since possible redundant elements that would result from the Boolean product operations do not simplify, this corresponding to a decreased overall probability.

E. Noncoherent Systems

In previous sections we have devoted our attention to s-coherent systems, i.e., in Boolean terms, to systems in which the top event, $\overline{\phi}(X) \ge \overline{\phi}(X')$ for $X_n \ge X_n'$ and all indices n. It is not so for noncoherent systems. This means that some bounding methods valid for s-coherent systems cannot be generally applied to noncoherent ones, in particular, the min-path and min-cut bound method. With the inclusion-exclusion method lower and upper bounds cannot be identified a priori, although two successive terms of the sums Equations 73 and 78 may be used to bracket the availability and unavailability, respectively.

In view of defining a nomenclature for the description also of noncoherent structures, the concepts of path sets and cut sets, relevant to s-coherent systems, are generalized into those of implicates and implicants, respectively. As with path sets and cut sets, they are given by Boolean intersections of elementary events, one of which may have a given component state indicator and another one its complement. As with minimal path sets and cut sets in s-coherent systems, the implicates and the implicants reduced by Boolean elimination of redundant elements are termed prime implicates and implicants, respectively.

With the above definitions of implicates and implicants, the inclusion-exclusion method (as far as bracketing values of the system unavailability are concerned) and the min-max method can be used for noncoherent systems as well if we introduce in the corresponding relationships their Boolean expressions in place of those relevant to the path sets and cut sets, respectively.

Noncoherent systems can be also represented by fault trees, if not gates are introduced, so called since they negate the event corresponding to the branch at the point considered, i.e., they have the effect of transforming its Boolean indicator (1 or 0) into its complement (0 or 1). In Figure 9A an example of a fault tree, with simplified symbology, is shown relevant to a noncoherent system.

In order to reduce the difficulties encountered in fault tree analysis when dealing with noncoherent systems, the given tree structure may be transformed into that of an equivalent one relevant to a coherent system. This may be accomplished in two steps. First, the not gates are lowered to the component level (i.e., to basic events). Second, the tree is decomposed into modules, so that the complementary events are grouped within one module. The structure of the equivalent tree comprising these modules is now coherent.[16]

The step of moving down the not gates to the component level does not pose particular difficulties. To show this, it suffices to recall that the complement of an event (not necessarily the top event) represented by a given fault tree branch is given by its dual, obtained by replacing the elementary events corresponding to it ($\overline{A}_n$ or A_n) with their complements (A_n or $\overline{A}_n$), and the and (or or) gates with the or (or and) ones, respectively. For illustration, consider the example shown in Figure 9A. The equivalent tree in which the not gates are lowered to the component level is shown in Figure 9B, while the tree reduced into modules Y_1, Y_2, and Y_3 and the (simpler to analyze) module structure in Figure 9C.

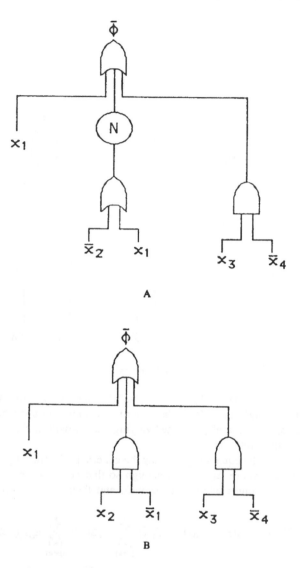

FIGURE 9. (A) Fault tree for noncoherent system, (B) fault tree with lowered not gate, and (C) reduced fault tree.

V. PROBABILISTIC ANALYSIS BY MARKOV CHAIN MODELS

For a given system formed by N components let us consider the complete set of its possible 2^N states, corresponding to all the possible combinations of its N binary (i.e., at up or down conditions) components.* If we assume that the transition probabilities from each state to the others do not depend on the past history of the system, we can identify these state transitions with a Markov chain. As is well known, a Markov chain is a random process the evolution of which is influenced only by the present situation and not by the nature of the past.

---

* For simplicity, we assume here for each component only two possible states (up or down). In case components are characterized by more than two states (as, e.g., with standby ststems) the number of possible system states is correspondingly increased.

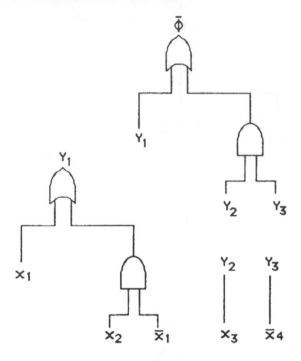

FIGURE 9C.

Let us assume that $P_k(t)$ [$k = 1,2, \ldots K(\equiv 2^N)$] is the probability that the system is in the kth state at time t and that $a_{kj}(t)$ is the conditional transition probability from state j to state k. Quantities a_{kj} are generally assumed varying very slowly, and can then be considered constant, for large time intervals.

In order to determine the equation governing the generic probability $P_k(t)$, we shall consider it at the time $t + \Delta t$, where Δt is small enough so that no multiple transitions occur in it (i.e., implying state changes of more than one), and then write a probability balance. It results

$$P_k(t + \Delta t) = P_k(t) - P_k(\bar{t}) \sum_{\substack{j=1 \\ (j \neq k)}}^{K} a_{jk}\Delta t + \sum_{\substack{j=1 \\ (j \neq k)}}^{K} a_{kj}P_j(\bar{t})\Delta t \qquad (101)$$

where $\bar{t}$ represents a time in the interval $(t, t + \Delta t)$.

If we define

$$a_{kk} = - \sum_{\substack{j=1 \\ (j \neq k)}}^{K} a_{jk} \qquad (102)$$

and rearrange Equation 101, we obtain

$$\frac{P_k(t + \Delta t) - P_k(t)}{\Delta t} = a_{kk}P_k(\bar{t}) + \sum_{\substack{j=1 \\ (j \neq k)}}^{K} a_{jk}P_k(\bar{t}) \qquad (103)$$

and, making $\Delta t \to 0$,

$$\frac{dP_k(t)}{dt} = a_{kk}P_k(t) + \sum_{\substack{j=1 \\ (j \neq k)}}^{K} a_{jk}P_k(t) \qquad (104)$$

If we define the transition matrix

$$
M_a = \begin{vmatrix}
a_{11} & a_{12} & \cdot & \cdot & \cdot & a_{1K} \\
a_{21} & a_{22} & \cdot & \cdot & \cdot & a_{2K} \\
\cdot & \cdot & \cdot & \cdot & \cdot & \cdot & , \\
\cdot & \cdot & \cdot & \cdot & \cdot & \cdot & \cdot \\
a_{K1} & a_{K2} & & & & a_{KK}
\end{vmatrix} \tag{105}
$$

and the state probability vector

$$
P(t) = \begin{vmatrix}
P_1(t) \\
P_2(t) \\
\cdot \\
\cdot \\
\cdot \\
P_K(t)
\end{vmatrix} \tag{106}
$$

Equation 104 above can be written

$$
\frac{dP(t)}{dt} = M_a P(t) \tag{107}
$$

Let us assume that state 1 corresponds to a success state with all components working. The initial condition of $P_k(t)$, at $t = 0$, is assumed

$$
P(0) = \begin{vmatrix}
1 \\
0 \\
\cdot \\
\cdot \\
\cdot \\
0
\end{vmatrix} \tag{108}
$$

In order to be integrated, Equation 107 is generally discretized in time intervals (t_1, t_{1+1}), small enough so that within each of them matrix M_a can be assumed constant. The solution for $P(t)$ can then be obtained by the recurrent form

$$
P(t_1) = P(t_{1-1})[1 + M_a(t_{1-1})(t_1 - t_{1-1})] \qquad (1 = 1,2,\ldots) \tag{109}
$$

The K ($\equiv 2^N$) states are usually ordered into two categories: that relevant to the so-called success (or up) states, i.e., states in which the system is in normal operation, and failure (or down) states, i.e., states in which the system is failed. Let us assume that the first I states correspond to success (up) ones. We can easily see that the system availability $A_s(t)$ can be defined as

$$
A_s(t) - \sum_{k=1}^{I} P_k(t) - 1 - \sum_{k=I+1}^{K} P_k(t) \tag{110}
$$

In order to determine the system reliability $R_s(t)$ the transition matrix has to be modified. Let us consider first the original matrix M_a partitioned as shown below

$$M_a = \begin{vmatrix} M_{11} & \vdots & M_{12} \\ -----&--&---- \\ M_{21} & \vdots & M_{22} \end{vmatrix} \tag{111}$$

where M_{11}, M_{12}, M_{21}, and M_{22} are submatrices corresponding to transitions from up to up, from down to up, from up to down, and from down to down states (assuming components can fail also with the system down), respectively. We define then matrix

$$\overline{M}_r = \begin{vmatrix} M_{11} & \vdots & 0 \\ -----&--&---- \\ M_{21} & \vdots & M_{22} \end{vmatrix} \tag{112}$$

obtained by eliminating all the transitions from down to up states. At this point we can consider all failed states merged into a single "absorption state", i.e., such that its probability $\hat{P}_{abs}$ is the sum of those corresponding to all the (K-I) down states, i.e.,

$$\hat{P}_{abs}(t) = \sum_{k=I+1}^{K} \hat{P}_k(t) \tag{113}$$

Summing the last (K-I) lines of expression, Equation 112, it is easy to verify that $\hat{P}_{abs}(t)$ obeys the equation

$$\frac{d\hat{P}_{abs}(t)}{dt} = \sum_{k=I+1}^{K} [a_{k1}\hat{P}_1(t) + a_{k2}\hat{P}_2(t) + \dots + a_{kI}\hat{P}_I(t)] \tag{114}$$

where state probabilities $\hat{P}_k$ generally differ from P_k. In place of the (K-I) down states we consider now this absorbing one and replace matrix $\overline{M}_r$, Equation 112, with

$$\hat{M}_r = \begin{vmatrix} M_{11} & \vdots & 0 \\ -----&--&---- \\ \sum_{k=I+1}^{K} a_{k1} \quad \sum_{k=I+1}^{K} a_{k2} \cdots \sum_{k=I+1}^{K} a_{kI} & \vdots & 0 \end{vmatrix} \tag{115}$$

The new state probability vector

$$\hat{P}(t) = \begin{vmatrix} \hat{P}_1(t) \\ \hat{P}_2(t) \\ \cdot \\ \cdot \\ \cdot \\ \hat{P}_I(t) \\ \hat{P}_{abs}(t) \end{vmatrix} \tag{116}$$

obeys then the equation

$$\frac{d\hat{P}(t)}{dt} = \hat{M}_r \hat{P}(t) \tag{117}$$

and the system reliability results

$$R_s(t) = \sum_{k=1}^{1} \hat{P}_k(t) = 1 - \hat{P}_{abs}(t) \tag{118}$$

We give in the following a few simple examples for illustration. Consider first a system consisting of a single component with constant failure and instantaneous repair rate λ and μ, respectively. For this system we can define two distinct states: state 1, with probability $P_1(t)$ in which the system is up, and state 2, with probability $P_2(t)$, in which the system is down. The probability of transition from state 1 to state 2 corresponds to the probability that the system undergoes a failure (i.e., transits into state 2) given it works (being in state 1). We know that this probability is equal to λ. Likewise, we can define transition from state 2 into state 1. We can then write the system

$$\frac{d}{dt} \begin{vmatrix} P_1 \\ P_2 \end{vmatrix} = \begin{vmatrix} -\lambda & \mu \\ \lambda & -\mu \end{vmatrix} \begin{vmatrix} P_1 \\ P_2 \end{vmatrix} \tag{119}$$

If we define its initial value

$$\begin{vmatrix} P_1 \\ P_2 \end{vmatrix} = \begin{vmatrix} 1 \\ 0 \end{vmatrix} \tag{120}$$

we easily obtain the solution

$$P_1(t) = \frac{1}{\lambda + \mu} [\mu + \lambda e^{-(\lambda + \mu)x}] \tag{121}$$

$$P_2(t) = \frac{\lambda}{\lambda + \mu} [1 - e^{-(\lambda + \mu)x}] \tag{122}$$

so that the system availability results $A_s(t) = P_1(t)$ and tends to $\mu/(\lambda + \mu)$ for $t \to \infty$.
For what concerns the system reliability, modifying the transition matrix into

$$\hat{M}_r = \begin{vmatrix} -\lambda & 0 \\ \lambda & 0 \end{vmatrix} \tag{123}$$

the expected exponential function $e^{-\lambda t}$ is obtained.
Consider now a system of two components in parallel of which only one is required for operation and characterized by constant instantaneous failure and repair rates λ_1, μ_1 and λ_2, μ_2, respectively. Only one repairman is assumed and no failed state while at standby. The possible states of this system are shown in the table below.

State	Component 1	Component 2	System
1	Up	Standby	Up
2	Standby	Up	Up
3	Under repair	Up	Up
4	Up	Under repair	Up
5	Under repair	Down	Down
6	Down	Under repair	Down

It is not difficult now to determine the transition probabilities between different states, considering the conditional transition probabilities from working (up) to failed (down) and

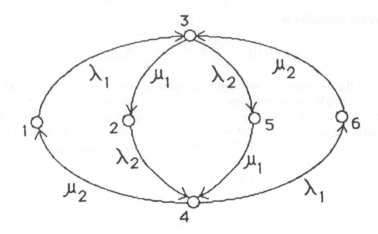

FIGURE 10.   Markov graph example.

to failed to working, or standby, conditions relevant to each component. Recalling that at operating conditions, once a component is repaired, this goes automatically at standby, the following transition matrix can then be written

$$
M_s =
\begin{vmatrix}
-\lambda_1 & 0 & 0 & \mu_2 & 0 & 0 \\
0 & -\lambda_2 & \mu_1 & 0 & 0 & 0 \\
\lambda_1 & 0 & -(\lambda_2 + \mu_1) & 0 & 0 & \mu_2 \\
0 & \lambda_2 & 0 & -(\lambda_1 + \mu_2) & \mu_1 & 0 \\
0 & 0 & \lambda_2 & 0 & -\mu_1 & 0 \\
0 & 0 & 0 & \lambda_1 & 0 & -\mu_2
\end{vmatrix}
\tag{124}
$$

For complex systems a Markov state transition graph can be helpful, as shown in Figure 10, relevant to the case considered above. We see how, in a Markov state transition graph, edges from left to right are indicative of failure transitions while edges from right to left are indicative of repair transitions.

If we assume that the two components are identical, then states 1,2, as well as states 3,4 and 5,6 can be merged, so that the following states can be defined

- State 1: one unit operating and the other one at standby (system up)
- State 2: one unit operating and the other one under repair (system up)
- State 3: one unit down and one under repair (system down)

The corresponding transition matrix will be (setting $\lambda = \lambda_1 = \lambda_2$ and $\mu = \mu_1 = \mu_2$)

$$
M_a =
\begin{vmatrix}
-\lambda & \mu & 0 \\
\lambda & -(\mu + \lambda) & \mu \\
0 & \lambda & -\mu
\end{vmatrix}
\tag{125}
$$

The corresponding equation governing $P(t)$ can be easily integrated (e.g., making use of Laplace transforms). The solution of P_2, in particular, gives the system unavailability $\overline{A}_a$ and, then, the system availability $A_a [= 1 - \overline{A}_a(t)]$. for $t \rightarrow \infty$ the steady-state availability results

$$
A_a(\infty) = 1 - P_2(\infty) = \frac{\mu + \mu\lambda}{\mu^2 + \mu\lambda + \lambda^2}
\tag{126}
$$

If we replace matrix M_a with matrix

$$\hat{M}_r = \begin{vmatrix} -\lambda & \mu & 0 \\ \lambda & -(\mu + \lambda) & 0 \\ 0 & \lambda & 0 \end{vmatrix} \tag{127}$$

the reliability $R_s(t)$ can, instead, be obtained. It results

$$R_s(t) = \frac{-s_2 e^{-s_1 t} + s_1 e^{-s_2 t}}{s_1 - s_2} \tag{128}$$

where

$$s_1 = \frac{-\mu - 2\lambda + \sqrt{\mu^2 - 4\mu\lambda}}{2} \tag{129}$$

$$s_2 = \frac{-\mu - 2\lambda - \sqrt{\mu^2 - 4\mu\lambda}}{2} \tag{130}$$

For simple systems with few components the solution of Equations 107 and 117 may be obtained analytically, as seen above. For realistic situations in which the systems consist of a large number (N) of components the task may, instead, become formidable, considering that, for two state components, the possible system states are 2^N. However, due to some simplifying characteristics encountered with many systems and in consideration of the considerable memory capacity and rapidity of calculation of modern computers, the Markov chain approach to the probabilistic analysis of systems is becoming a viable methodology in system performance and safety studies.

As we have seen in the derivation of Equation 107, governing the state probability vector $P(t)$, multiple transitions are excluded, i.e., only transitions from a state with, e.g., 1 (out of N) failed components to states with $1'$ failed components are allowed, such that $|1 - 1'| = 1$. If we consider the same matrix partition as shown in Equation 111, ordinate the system up states according to the number 1 [1 = 1,2, . . . L (<N)] of failed components and do the same for the down states, the resulting matrices $M_{11}, M_{12}, M_{21},$ and M_{22} take the form[17]

$$M_{11} = \begin{vmatrix} M_{11}^{00} & M_{11}^{01} & 0 & 0 & \cdots & 0 \\ M_{11}^{10} & M_{11}^{11} & M_{11}^{12} & 0 & \cdots & 0 \\ 0 & M_{11}^{21} & M_{11}^{22} & M_{11}^{23} & \cdots & 0 \\ \cdots\cdots\cdots\cdots\cdots\cdots\cdots\cdots\cdots\cdots \\ \cdots\cdots\cdots\cdots\cdots\cdots\cdots\cdots\cdots\cdots \\ 0 & 0 & 0 & 0 & \cdots & M_{11}^{LL} \end{vmatrix} \tag{131}$$

$$M_{12} = \begin{vmatrix} M_{12}^{01} & 0 & 0 & \cdots & 0 & \cdots \\ 0 & M_{12}^{12} & 0 & \cdots & 0 & \cdots \\ \cdots\cdots\cdots\cdots\cdots\cdots\cdots\cdots\cdots\cdots \\ \cdots\cdots\cdots\cdots\cdots\cdots\cdots\cdots\cdots\cdots \\ 0 & 0 & 0 & \cdots & M_{12}^{L,L+1} & \cdots & 0 \end{vmatrix} \tag{132}$$

$$M_{21} = \begin{vmatrix} M_{21}^{10} & 0 & 0 & \cdots & 0 \\ 0 & M_{21}^{21} & 0 & \cdots & 0 \\ \cdots\cdots\cdots\cdots\cdots\cdots\cdots\cdots\cdots\cdots\cdots \\ \cdots\cdots\cdots\cdots\cdots\cdots\cdots\cdots\cdots\cdots\cdots \\ 0 & 0 & 0 & \cdots & M_{21}^{L+1,L} \\ \cdots\cdots\cdots\cdots\cdots\cdots\cdots\cdots\cdots\cdots\cdots \\ \cdots\cdots\cdots\cdots\cdots\cdots\cdots\cdots\cdots\cdots\cdots \\ 0 & 0 & 0 & \cdots & 0 \end{vmatrix} \tag{133}$$

$$M_{22} = \begin{vmatrix} M_{22}^{11} & M_{22}^{12} & 0 & 0 & \cdots & 0 \\ M_{22}^{21} & M_{22}^{22} & M_{22}^{23} & 0 & \cdots & 0 \\ 0 & M_{22}^{32} & M_{22}^{33} & M_{22}^{34} & \cdots & 0 \\ \cdots\cdots\cdots\cdots\cdots\cdots\cdots\cdots\cdots\cdots\cdots \\ \cdots\cdots\cdots\cdots\cdots\cdots\cdots\cdots\cdots\cdots\cdots \\ 0 & 0 & 0 & 0 & & M_{22}^{NN} \end{vmatrix} \tag{134}$$

where $M_{\alpha\beta}^{11'}$ ($\alpha,\beta = 1,2$) represent submatrices relevant to transitions from states with $1'$ into states with 1 failed components. Submatrices $M_{\alpha\alpha}^{11}$ are diagonal.

When so ordered, Equation 107 can then be solved with significant reduced effort.

A considerable simplification of the problem can be obtained if the Markov processes describing the system evolution exhibit certain symmetries, which is frequently encountered in complex highly redundant systems. In these cases the dimensions of matrix M_a can be reduced by merging blocks of system states into superstates (each associated with a probability given by the sum of those relevant to the states included in the corresponding block). These symmetries, both at component and subsystem level, correspond to the possibility of inter-changing two elements (with the same instantaneous failure and repair rates), or two sub-systems (with one to one correspondence between components, each of these having the same instantaneous failure and repair rates), without altering the success, or failure, state condition.[18,19]

A further method aimed at simplifying the Markov chain structure will be described in Appendix B of Chapter 5 in the derivation of which use is made of sensitivity method concepts.

As far as numerical methods solving Equation 107 are concerned, existing computational codes developed in different fields for similar types of problems can be used, e.g., the well-established ORIGEN Code,[20] developed for nuclide transmutation calculations and using the matrix exponential method.*

* This method corresponds to expanding the formal solution, $\exp(M_a t)P(0)$, where $P(0) = P(t = 0)$, of Equation 107, i.e.,

$$P(t) = \left[1 + M_a t + \frac{(M_a t)^2}{2!} + \dots \right] P(0)$$

For relatively simple cases, the well known Laplace transform method can be used.*

VI. MONTE CARLO METHODS

Monte Carlo (MC) techniques are often adopted in probabilistic studies relevant to large complex systems.[21,22] The MC approach to availability and reliability analysis is in particular indicated where the statistical data related to the behavior of the components is highly uncertain and incomplete. The appropriate sampling over the distribution, properly taking into account the dispersion of the data, is in this case mandatory. MC techniques are also appropriate when the reliability block diagram relevant to a given system is complicated either due to large path sets or due to complex logical conditions in the definitions of a system failed state.[23]

Given a system formed by N components, the MC method basically consists in generating by simulation techniques a sample consisting in a number (e.g., M) of s-independent random values b_m of the component state vector $X = |X_1 \, X_2...X_N|^T$, starting from their respective distribution functions, so that at each vector b_m a value (1 or 0) of the system state structure function $\phi(X)$ can be associated. The system availability estimator obtained from the sample will then result

$$\tilde{A}_s = \frac{1}{M} \sum_{m=1}^{M} \Phi(b_m) \tag{135}$$

The variance $[\text{var}(\tilde{A})]$ associated with $\tilde{A}$, by which confidence intervals can be determined, can be obtained from elementary considerations, as shown in the following.

Let us consider the quantity $\tilde{y} = M\tilde{A}_s$ $(<M)$. Since $\tilde{y}$ corresponds to the number of (system) successes out of a sample of M trials, it has a binomial distribution. Using symbol y to indicate the possible values of the estimator, we can then write

$$P(y = r) = \binom{N}{r}\pi^r(1 - \pi)^{M-r} \tag{136}$$

π being the probability that the outcome of a trial gives a success and then corresponds exactly to the availability A_s. The variance of $\tilde{A}_s$ then results

$$\text{var}(\tilde{A}_s) \approx \text{var}(y/M) = \frac{1}{M^2} \left[\sum_{r=0}^{M} r^2 P(y = r) - (MA_s)^2 \right]$$

$$= \frac{1}{M^2} \left[\sum_{r=2}^{M} r(r - 1)P(y = r) + \sum_{r=1}^{M} rP(y = r) - (MA_s)^2 \right]$$

---

* By this method, Equation 107 is transformed into equation

$$(sI - M_s)L[P(t)] = P(0)$$

L[P(t)] being the Laplace transform of P(t), I the unit matrix, and P(0) the initial conditions, [usually, P(0) = |1 0 . . . 0|]. Having solved for the transform L[P(t)], the solution for P(t) is then obtained, via antitransformation, i.e.,

$$P(t) = L^{-1}\left[\frac{\text{adj}(sI - M_s)}{\text{det}(sI - M_s)}\right]P(0)$$

adj(A) being the transposed matrix of cofactors of A.

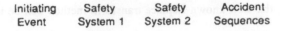

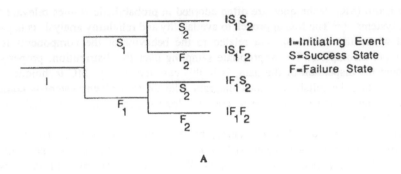

A

FIGURE 11.   (A) Event tree example and (B) event tree simplified, due to F_1 implying F_2 in example, Figure 11A.

$$= \frac{1}{M^2} [M(M - 1)A_s^2 + MA_s - (MA_s)^2]$$

$$= \frac{1}{M} A_s(1 - A_s) \tag{137}$$

and can be estimated replacing A_s with $\tilde{A}_s$, given by Equation 135.

An analogous procedure could be followed for estimating the system reliability R_s and its variance.

The procedure illustrated above corresponds to what is called the crude MC method. The presently available MC codes, developed for system probabilistic analysis either using fault tree techniques or the Markov chain approach, are in fact characterized by sophisticated procedures aimed at reducing the variance, given the same calculational effort, or, which is the same, speeding up the calculation for the same accuracy.[24-26]

VII. EVENT TREE

The event tree represents an inductive logical diagram. Starting from an initiating accidental event and following subsequent bifurcations [in relation to binary (up, down) subsystems, otherwise more than two branches should be considered], it describes various sequences of events corresponding to as many system states, each associated with a particular consequence.

The event tree methodology under certain aspects resembles that relevant to the decision tree in business applications. It represents the necessary framework for the overall risk probability assessment (RPA) by providing a basis in defining accident scenarios in relation to each initiating (internal or external) accidental event and to the success, or failure, of the safety related systems. Besides, its use allows to define top events for system fault tree analysis.

For illustration, in Figure 11A the structure of a simple event tree is shown relevant to two safety systems involved after the occurrence of an initiating event. For more complex systems larger event trees will be obtained by simply continuing this stepwise branching.

Event trees can become very complicated, producing an enormous number of sequences to be considered. However, the presence of functional interdependences helps to eliminate

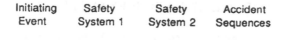

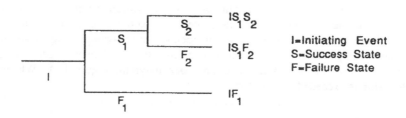

I=Initiating Event
S=Success State
F=Failure State

FIGURE 11B.

many of these sequences. For instance, if in the example shown (Figure 11A) the failure of system 1 has produced that of system 2, the diagram simplifies, as indicated in Figure 11B.

Once the event tree has been constructed, each sequence of events is associated to a risk category, depending upon the gravity of its evaluated consequences.

To each fault event F, a (conditioned) probability is usually assigned, using, when necessary, the fault tree (or other adequate) analysis techniques. Quantitative evaluations can then be obtained of the probability associated to each sequence.(*)

APPENDIX A: BOOLEAN ALGEBRA

The Boolean algebra can be applied to the elements of a set, each characterized by one of two possible states (events), subject to binary union ($\cup$) and intersection ($\cap$) operations having as result elements of that same set. Such two states are identified themselves as elements of the set as "universal event" and "null event", corresponding to the binomial 1-0 (representing binomials of the type true-false, up-down [working-failed], etc., depending on the class of elements considered), respectively.

The union and intersection operations of events X_n ($n = 1, 2, ... N$) (each with value either 1 or 0, depending on the state of the corresponding element) satisfy the following relationshpis (generically indicating with "$\circ$" symbols $\cup$ or $\cap$, and with "σ" symbols $\cap$ or $\cup$, respectively):

- $X \circ Y = Y \circ X$ (commutative law)
- $(X \circ Y) \circ Z = X \circ (Y \circ Z)$ (associative law)
- $X \circ X = X$ (idempotent law)
- $X \circ (X \sigma X) = X$ (absorption law)
- $X \circ (Y \sigma Z) = (X \circ Y) \sigma (X \circ Z)$ (distributive law)
- $X \cup X = 1, X \cap \overline{X} = 0, \overline{(\overline{X})} = X$ (complementation laws)
- $\overline{X \circ Y} = \overline{X} \sigma \overline{Y}$ (De Morgan's law)
- $X \cup 1 = 1, X \cap 1 = X$
- $X \cup 0 = X, X \cap 0 = 0$

In practical cases, which may imply a large number of elements, the union and intersection symbols are replaced for simplicity with the standard ones of sum ($+$) and product ($\cdot$, often omitted). For example, $(X + Y)Z$ is used in place of $(X \cup Y) \cap Z$.

* For example, to the sequence IF_1F_2 in the event tree shown at Figure 11A the joint probability $P(IF_1F_2) = P(I)P(F_1|I)P(F_2|IF_1)$ can be associated. If there is full correlation between events F_1 and F_2 then $P(F_2|IF_1) = 1$ and, consequently, $P(IF_1F_2) = P(IF_1)$, which justifies in this case the simplified tree of Figure 11B.

APPENDIX B: ILLUSTRATIVE EXAMPLE

To further illustrate the methodology illustrated in Section 3, the fault and success trees will be here described relevant to a (simplified) temperature fast shutdown (scram) system of a nuclear reactor project.[27] This system consists of a protective function, an output logic and an electromechanical subsystem. The protective function consists of a chain of three redundant channels (each made up by an independent thermocouple, signal amplificator, and threshold comparator), each of which feeds into three redundant electronic lines, as schematically shown in the system function block diagram, Figure 12A. Whenever the threshold value is exceeded in a protective function channel, this enters into a trip state. If more than one of the three channels are in a trip state, the signal in the three main electronic lines is interrupted. Each control rod is supported by a magnet made of three totally independent coils. Each coil is electrically fed by a power switchgear supplied by the ASR module directly connected to one of the three main electronic lines; if there is no signal in at least two out of three main lines, the two switchgears in each of the magnets are opened and, since one coil is not enough to support a rod, all of them fall by gravity and the reactor shutdown is activated.

To simplify the analysis of the system, we define the following Boolean indicators:

- Sn = 1, or 0, if the thermocouple n (n = 1,2,3) is operating (up), or failed to danger (down), respectively
- An = 1, or 0, if the amplifier n (n = 1,2,3) is up or down
- Tn = 1, or 0, if the treshold comparator n (n = 1,2,3) is up or down
- Dn = 1, or 0, if the decoupler n (n = 1,2,3) is up or down
- Mn = 1, or 0, if the 2/3 module n (n = 1,2,3) is up or down
- Nn = 1, or 0, if the and module n (n = 1,2,3) is up or down
- Rn = 1, or 0, if the ASR module n (n = 1,2,3) is up or down
- ϕ (structure function) = 1, or 0, if scram intervenes when required, or not, respectively
- $\bar{\phi}$ (complemented structure function, or Top Event) = 1, or 0, if scram does not intervene when required, or it does, respectively
- $\bar{X}$ (complemented component indicator) = 1, or 0, if component is failed to danger (i.e., with failure contributing to top event), or not, respectively.

In order to inductively construct the fault tree relevant to the complemented structure function (top event) $\bar{\phi}$, the successive subordinate failure events that may contribute to its occurrence must be first identified and linked to it by logical connective functions. The subordinate events themselves are then broken down to their logical contributions until the branching is terminated with primary fault events. Following these guidelines, the fault tree shown in Figure 12B can be constructed.

Function $\bar{\phi}$ can be expressed analytically using Boolean algebra: the following reduced expression can be obtained:

$$\bar{\Phi} = (\bar{S}1 + \bar{A}1 + \bar{T}1 + \bar{D}1)(\bar{S}2 + \bar{A}2 + \bar{T}2 + \bar{D}2 + \bar{S}3 + \bar{A}3 + \bar{T}3 + \bar{D}3)$$
$$+ (\bar{S}2 + \bar{A}2 + \bar{T}2 + \bar{D}2)(\bar{S}3 + \bar{A}3 + \bar{T}3 + \bar{D}3)$$
$$+ (\bar{M}1 + \bar{N}1 + \bar{R}1)(\bar{M}2 + \bar{N}2 + \bar{R}2 + \bar{M}3 + \bar{N}3 + \bar{R}3)$$
$$+ (\bar{M}2 + \bar{N}2 + \bar{R}2)(\bar{M}3 + \bar{N}3 + \bar{R}3)$$

from which the various minimal cut sets (here, pairs of fault events) can be easily identified.

To the above expression of the top event the reduced fault tree of Figure 12C can be associated. The corresponding success tree shown in Figure 12D is obtained immediately

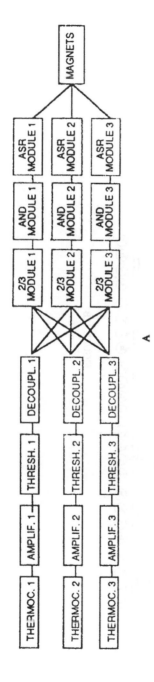

FIGURE 12.   (A) Function block diagram of shutdown system, (B) fault tree, (C) reduced fault tree, (D) success tree, and (E) logic block diagram.

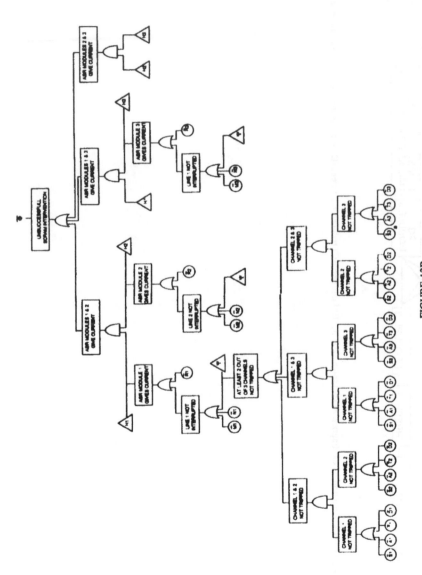

FIGURE 12B.

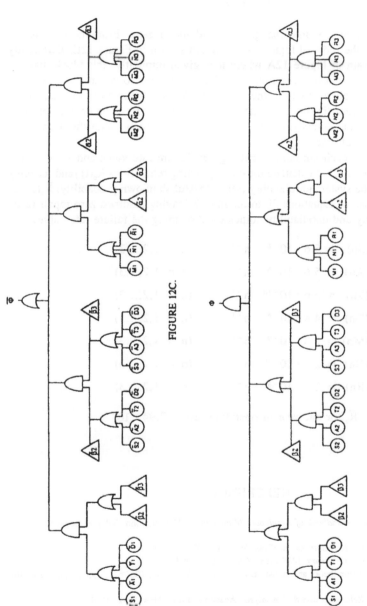

FIGURE 12C.

FIGURE 12D.

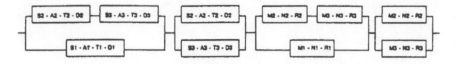

FIGURE 12E.

by complementing the events and interchanging the and and or gates. From the success tree it is also easy to obtain the reduced logic block diagram shown in Figure 12E. Differently from the function diagram of Figure 12A where it is given implicitly, this block diagram explicitly displays the two-out-of-three logic. A (deductive) way to construct a fault tree could then be suggested: obtain first (in the case considered starting from the scheme of Figure 12A) a logic diagram (in this case that given in Figure 12E, or an equivalent one), then from this construct the corresponding success tree and, finally, with the usual complementation rules, the fault one.

If we associate now to each nth component a given failure rate value and assume zero repair rates, using Equation 21 to calculate the corresponding reliabilities $R_n(t)$ [and recalling that $\overline{R}_n = (1 - R_n)$], the system unreliability (unsuccessful shutdown probability) $\overline{R}_s(t)$ can be calculated making use of Equations 78 to 82 and 67 (having assumed zero repair rates, the system unavailability and unreliability coincide). Assuming the failure rate values:

$$\lambda(Sn) = 5.6 \cdot 10^{-6} \quad h^{-1} \qquad (n = 1,2,...3)$$

$$\lambda(An) = 5.6 \cdot 10^{-6} \quad h^{-1} \qquad (n = 1,2,...3)$$

$$\lambda(Tn) = 5.6 \cdot 10^{-10} \quad h^{-1} \qquad (n = 1,2,...3)$$

$$\lambda(Dn) = 4.4 \cdot 10^{-6} \quad h^{-1} \qquad (n = 1,2,...3)$$

$$\lambda(Mn) = 2.8 \cdot 10^{-9} \quad h^{-1} \qquad (n = 1,2,...3)$$

$$\lambda(Nn) = 2 \quad \cdot 10^{-10} \quad h^{-1} \qquad (n = 1,2,...3)$$

$$\lambda(Rn) = 2 \quad \cdot 10^{-8} \quad h^{-1} \qquad (n = 1,2,...3)$$

the system unreliability $\overline{R}_s$ for one year of operation results $7.94 \cdot 10^{-3}$.

REFERENCES

1. **Ledermann, W., Ed.,** *Handbook of Applicable Mathematics,* Vol. 6, John Wiley & Sons, Chichester, England, 1983.
2. **Ireson, W. G.,** *Reliability Handbook,* McGraw-Hill, New York, 1958.
3. **Raisbeck, G.,** *Information Theory,* MIT Press, Cambridge, 1963.
4. **Jaynes, E. T.,** Information theory and statistical mechanics, *Phys. Rev.,* 106, 620, 1957; *Phys. Rev.,* 108, 171, 1957
5. **McCormick, N. J.,** *Reliability and Risk Analysis,* Academic Press, New York, 1981.
6. **Chu, T. L. and Apostolakis, G. E.,** Methods for probabilistic analysis of noncoherent fault trees, *IEEE Trans. Reliab.,* 29, 354, 1980.
7. **Jackson, P. S.,** On the importance of elements and prime implicants on non-coherent systems, *IEEE Trans. Reliab.,* 32, 21, 1983.
8. **Zhang, Q. and Mei, Q.,** Element importance and system failure frequency of a 2-state system, *IEEE Trans. Reliab.,* 24, 308, 1985.
9. **Wu, J. S., Salem, S. L., and Apostolakis, G. E.,** The use of decision tables in the systematic construction of fault trees, in *Nuclear Systems Reliability Engineering and Risk Assessment,* Fussel, J. B. and Burdick, G. R., Eds., SIAM, Philadelphia, 1977.

10. **Tribus, M.,** *Rational Descriptions, Decisions and Design,* Pergamon Press, Oxford, 1969.
11. **Lambert, H. E.,** Fault Trees for Decision Making in System Analysis, Report UCRL-51829, Lawrence Livermore Laboratory, Livermore, CA, 1975.
12. **Garibba, S., Mussio, P., Naldi, F., Reina, G., and Volta, G.,** Efficient construction of minimal cut sets from fault trees, *IEEE Trans. Reliab.,* 26, 88, 1977.
13. **Jasmon, G. B. and Kai, O. S.,** A new technique in minimal path and cutset evaluation, *IEE Trans. Reliab.,* 34, 136, 1985.
14. **Fussel, J. B. and Vesely, W. E.,** A new methodology for obtaining cut sets for fault trees, *Trans. Am. Nuclear Soc.,* 15, 262, 1972.
15. **Barlow, R. E. and Proschan, F.,** *Statistical Theory of Reliability and Life Testing,* Holt, Reinhart & Winston, New York, 1975.
16. **Kumamoto, H. and Henley, E. J.,** Top-down algorithm for obtaining prime implicant sets of non-coherent fault trees, *IEEE Trans. Reliab.,* 27, 242, 1978.
17. **Papazoglou, I. A. and Gyftopoulos E. P.,** Markov processes for reliability analyses of large systems, *IEEE Trans. Reliab.,* 26, 232, 1977.
18. **Bacon, G. C.,** The decomposition of stochastic automats, *Inform. Contr.,* 7, 320, 1964.
19. **Kemeny, J. G. and Snell, J. L.,** *Finite Markov Chains,* D Van Nostrand, New York, 1960.
20. **Bell, M. J.,** ORIGEN, The ORNL Generation and Depletion Code, Report ORNL-4628, Oak Ridge National Laboratory, Oak Ridge, TN, 1973.
21. **Kamat, S. J. and Riley, M. W.,** Determination of reliability using event-based Monte Carlo simulation, *IEEE Trans. Reliab.,* 24, 73, 1975.
22. **Becker, P. W.,** Finding the better of two similar designs by Monte Carlo techniques, *IEEE Trans. Reliab.,* 23, 242, 1974.
23. **Dubi, A., Goldfeld, A., Leibowitz, A., and Sasson, D.,** Aspects of evaluation of complexed systems by the Monte Carlo method, Report MCU-1, Ben Gurion University, Beer Sheva, 1986.
24. **Kumamoto, H., Tanaka, K., and Inoue, K.,** Efficient evaluation of system reliability by Monte Carlo method, *IEEE Trans. Reliab.,* 26, 311, 1977.
25. **Mazumdar, M.,** Importance sampling in reliability estimation, in *Reliability and Fault Tree Analysis,* SIAM, Philadelphia, 1975, 153.
26. **Goldfeld, A. and Dubi, A.,** Montecarlo Methods in the Computation of System Reliability, *Energia Nucleare,* 4(3), 41, 1987.
27. **Righini, E.,** Description and reliability analysis of the PEC reactor shut-down system, in Proc. IAEA-NPPCI Specialists' Meeting on Actuating Devices in Control Systems for Nuclear Power Plants, Florence, December 1 to 3, 1986.

Chapter 4

RISK ANALYSIS AS AN INSTRUMENT OF DESIGN

Remo Galvagni and Stefano Clementel

TABLE OF CONTENTS

I. INTRODUCTION

Methodologies of risk analysis and assessment have been rapidly developing during the last decades, as a consequence of the need of assessing "objective" procedures able to permit quantitative risk evaluations, to be used as decision tools. The complexity of processes and systems presently adopted in several "high-risk" (actually much more often high-hazards) plants, together with the need of an answer to the question of the evaluation of the risk associated to their operation acceptable to the public, prevent the adoption of subjective methods only based on individual experience. Methods and theories have been developed and proposed for risk analysis, together with computing instruments which have greatly benefited by the fast development of cybernetics.

The research is still on-going; several approaches are presently available, and verifying aspects may be recognized.

In the following, after a short introduction on basic concepts, a picture will be provided of some main theories, explaining the conceptual phases of their development. Due to the nature of the matter treated, a philosophical approach has been adopted. Starting from Section IX.B, a more pragmatic approach has been used, allowing the exemplification of the main phases of a complete risk analysis methodology. Examples on different applications are also included.

II. DEFINITIONS

Concepts as safety, risk, reliability, availability, qualification, and quality are generally considered interconnected and dependent on one another. However, to transform this correlation from an implicit sensation into an explicit quantification, endowed, above all, with a validity as far as a possible objective, is a very hard enterprise and, up to now, it is very far from being fully accomplished.

First of all, let us define some of the concepts mentioned above, so that the meaning of what follows is rendered univocal, avoiding possible misunderstandings due to the different habits of those who operate in analogous fields but in different sectors.

Safety and risk vs. a particular damage are here used as complementary concepts, such as availability and nonavailability; while generically speaking, and after the common use, safety represents the complement to risk only when by risk a damage to health or "health damage" is meant.

Risk is defined as the estimated chance of incurring a certain damage or, more generally, as a magnitude furnishing the synthetic evaluation of a situation represented by the values of two variables: the entity of damage and its probability, i.e., by definition, a magnitude function of two variables one of which is probability.

Consequently the term "risk analysis" and not "probabilistic" risk analysis will be used; if a differentiation must be made it will be made between risk analysis and quantitative risk analysis.

Speaking of risk analysis will never intend consideration of only the health damage; in such a case, in fact, the term "safety analysis" will be used, while the damage considered in risk evaluation will always include the economic damage also, as that due to the non-availabilities of a plant and to the costs of maintenance.

By quality, instead, we mean the sum of the requirements of reliability, maintenance, availability, operation, and costs, which must be guaranteed in order to satisfy the risk aims assumed for the investment (plant). Qualification or requalification tests are defined as those aimed at demonstrating that, in the forecasted operating conditions, the component has, or maintains, the characteristics required by the specifications defining its quality (quality specifications) according to the given definition. Due to the frequent references to "event tree" and to "fault tree" in what follows, synthetic definitions of these are here provided (see also Section II, Chapter 1). Event tree is an inductive logical pattern which, starting from an initiatory event, shows all its possible scenarios of consequences. The event progression in each scene represents one of the event sequences which may follow the starting event. The construction of event trees constitutes the main stage of risk analysis because the fundamental logical relations conditioning the following quantitative evaluations are established in this phase. Substantially event tree is nothing other than the "decision tree" of the decisional analysis theory.

"Fault tree" is a logical pattern of the various damage combinations, in series or in parallel, which may lead to a predefined event, called "top". Faults are possible events, therefore a fault tree is the graphic description of the logic interrelationship of the single events which cause the top, through the so-called logic gates (and, or, inhibit, etc.) conditioning the uprising or the blockage of the fault logic up the tree towards the top.

As its function clearly shows, fault tree is an essentially deductive instrument and deductive is the logic it represents.

III. RISK ANALYSIS

The quantification of the established results of a design as well as the assessment *a posteriori* of the results achieved requires the explicitation, in terms of risk aims, of its goals. All this is neither usual nor generally accepted, but it is a reality which has become more and more evident to the operators in the safety field. On the other hand and, evidently not by chance, fundamental elements for the assessment of the results achieved, such as reliability, safety, availability, and maintenance, are the stated aims of risk analysis. Quantitative risk analysis is precisely the instrument for an explicit and demonstrable verification of the correspondence of the design decisions to the established enterprise goals. It is evident that the development of a design requires a series of decisional choices at more and more detailed levels, each level being fatally characterized by an imperfect knowledge of who makes it (a decision presupposes uncertainty: uncertainty presupposes a lack of knowledge). How much a limited knowledge may influence the final result and what increase of risk may ensue is a problem to be solved each time; only the effort of making those questions explicit and objective allows, especially in the presence of complex technologies or innovative situations, the control of the "realization" process. Obviously, such control requires constant vigilance not only of the implications of what one knows, of what one is aware of not knowing, but also of the risk of what one ignores. For that purpose, the systematic endeavour to widen and generalize the knowledge derived from experience, e.g., in the activities of the plant accidental event banks or of the fault data banks, is essential. Obviously these banks must not constitute simple gathering of numbers, but rather they must be rational collections of experimental data to refer to in the different cases, to compare, and which can offer, therefore, all the elements of knowledge necessary for a comparison.

To be useful in the control of a complex planning process, the instrument of quantitative risk analysis must satisfy certain conditions of reasonable management, economy, and, above all, promptness of response, promptness which permits an efficacious feed-back on the decision taken and guarantees the operators and the whole activity against sterile and late criticism, which is not finalized, or able to be finalized, to a recovery of the established enterprise goals.

Once the conditions of reasonable practicability, reasonable costs, and satisfactory promptness of response are satisfied, risk analysis must become an instrument of constant verification of all the decisions taken during the whole process of progressive detailing and freezing of the realization specifications. Therefore, it becomes the assessment instrument of correctness, complete and congruent of the said specifications, at plant, system, subsystem, component, design, construction, testing, operation, and maintenance levels.

Since risk is defined by event probability and related consequences, naturally risk analysis implies an identification of all the events which are possible and therefore which constitute the objects of the analysis, the evaluation of their probabilities, and the determination of their consequences. Techniques of logical probabilistic analyses and techniques of phenomenological analyses are, therefore, both essential to every method of risk analysis.

Incidentally, it must be said that exactly the same consideration is obviously also applied to the methods of safety analysis; the contrast which some would like to see between the so-called "deterministic safety" techniques and the "probabilistic safety" techniques is, therefore, quite amazing. When all the possible events of interest are defined *a priori* and when the event probabilities are preestablished, even if only in a qualitative but satisfactory way, the recourse to logical probabilistic analysis is no longer necessary and therefore one has only the problem of the mechanistic determination of the consequences. In most cases this is a practical and proper simplification, at least while one is in the field of known events or of events easily derived from them, but it must not be taken as a general theory.

In this connection, we want to stress that, granted, the necessity of the contents of logical probabilistic information, the richness of such content, and consequently the knowledge with which to undertake the decisional process are strictly conditioned by the power of the available analysis instruments.

The lack of ductile, quick, and effectual instruments must lead to research into their development and not to a theory of their uselessness. Moreover, always in the same connection, it is necessary to return, for a moment, to the ancient and metaphysical problem of real probabilities. The hopeless research for this kind of knowledge, the so-called objective knowledge, hides the real problem that is the formulation of coherent estimates of the margins of uncertainty which — on the basis of the knowledge available for the evaluatin — characterizes, at the decisional stage, the judgment of adequacy of the objects there involved.

Not to remain in an abstract field, if we want to judge the adequacy of a particular structure, we must evaluate the margin of uncertainty of the assumption that it will successfully work maintaining its function, up to the end of its service life, such a margin being a consequence of the indeterminateness of loads, of the uncertainty about the entity and the spatial distribution of stresses, of the uncertainty about the characteristics of the materials, etc.

Faced with this problem, the fact that some other structures, in presumably similar conditions, have been successful may be considered of no great importance, unless it constitutes an increase in the basic available knowledge, that is, unless an evaluation of the information is possible which can give weight to the above term "presumably similar". That, however, implies a submission of the whole available knowledge to criticism in order to arrange in it the experimental information obtained and to evaluate its real importance.

It is fundamental not to yield to the temptation of taking as a law of nature the observation that the result is indifferent in the absence of an evident cause conditioning it; ideally the principle of "indifference" may be considered of general validity only when the specific causes are really absent and not when their occurrence is merely ignored. Thus, until the contrary is proved, the real objective knowledge derived from the repeated experiment, e.g., of dice-casting, is not that the probability of each face is 1/6, but rather that evident specific result conditioning causes are absent and that, applying the principle of indifference, no face can behave differently from the others.

As rightly recalled by De Finetti in his work on the Probability Theory, it is possible to assign a probability to an event, provided that it can be univocally defined as the object of an ideal wager; it must be possible to verify unequivocally *a posteriori* through an experimental procedure whether it took place or not.

However, probability does not possess such characteristics; one cannot define the probability of a probability for the lack of an instrument able to verify the result of the "wager".

Not to risk many illogicalities, this innate peculiarity of probability must be accepted; the research, therefore, of real probability, objectively measureable, without reference to the degree of knowledge conditioning and defining it, is the greatest illogicality of all.

Once this metaphysical research is given up, it is easier to formulate — on the basis of a clearly qualified degree of knowledge — responsible and motivated judgments, quantifying the uncertainty inherent to the sum of the particular and general knowledge available. It is also easier allowing others to go along again through the logical processes followed, in order to verify and update the results obtained in the light of a new and wider knowledge. Without the illustrated motivation of the Bayesian approach, which will be described in the following and which is considered fundamental for a correct and balanced expression of reliability goals in terms of specifications, the undertaken way could seem arbitrary or simplistic; on the contrary we consider it, for intrinsic logical reasons, almost compulsory.

The transition from risk analysis to risk assessment takes place in the utilization of the results obtained not only within the context of the control of the goals of a well-determined activity (e.g., see the comparison of similar situations in order to detect the most favorable, or the design control activities here described), but with wider aims such as the comparison, for decisional purposes generally of political character, of different contexts or activities.

IV. METHODOLOGIES FOR ANALYSIS

An outline, even though brief, will be made on problems and methodologies relevant to logical probabilistic analysis, as they appeared in their progress, using the point of view of who was engaged in activities of development and application, in a decisional field, of the here-discussed techniques, from the early 1960s up to today. Obviously the following exposition, as well as the influence of the vision of a person still working to develop the methodologies and applications of risk analysis along well-determined lines, will be compulsorily schematic, in that it tries to radicalize the concepts in order to intensify the fundamental differences rather than the many affinities among the ways of approach presently adopted.

The logical part of the logical probabilistic analysis is that it identifies the field or, better, the object of the probabilistic analysis, differentiating what is possible from what is certain and what is impossible. Such differentiation, in an initial phase, takes place in a mainly intuitive way, with much categorization and with recourse to quite rough, methodological instruments. For example, with the aim of judging the suitability of safety shut-down devices and of verifying their dimensioning, it is necessary to analyze the possibility and the effectiveness of their intervention following malfunctioning or plant accidents. Then, in such an initial phase, the whole of the events examined, in that they have been judged possible, is quite small; the apparently minor malfunctions are all excluded; the possibility of unsuccessful multiple interventions or concomitance of malfunctions is also excluded by recourse to a qualitative and often largely arbitrary probability estimation, based on considering them as incredible.

The substantial lack of methodological instruments used pushes towards a definition of events, identified by recourse to experience or to its extrapolation, to intuition, and to the analyst's imagination. Because of the reduced number of events considered and of the schematization of such events according to conditions of malfunctioning and assumptions thought to be conservative and often on the basis of not very rigorous elements, the probabilistic evaluations lose their clarity to evaporate into judgments such as "extreme caution" or "incredibility". On the other hand, the dimensioning verifications or, as they are usually defined, the capability assessment is emphasized, carried out with reference to situations assumed as archetypes and as such considered indisputably "true". Nevertheless, merit of this first phase is that it brings, though progressively and not without recourse to undersirable experiences, to the definition of classes of events, conventionally defined and better arranged and categorized which, through standardization of typologies of consequences, permit the evolution into a second, more analytical phase.

In this second phase we start from the events representative of the relative typologies of consequences to systemically and deductively analyze their modes of occurrence. In the course of this undertaking, it often becomes evident that it is necessary to discriminate among "top" events previously considered undifferentiated and to create more rigorous rules to individuate the differentiations. It is in this phase that, through the use of methodologies of analysis of the inductive type (FMEA, operability, etc.), we try to get over, though only partially, the subjectivity of the intuitive process of the previous phase, enabling the analyst to clarify at least the fundamental steps in his reasoning and so to "uncover" at least one part of his "true" knowledge (the comprehensiveness of understanding).

While the concept of true probability is refused, the essential concept of the true state of knowledge is emphasized.

Instruments and methodologies used in this second phase are those characteristic of a logical effort of inductive investigation which is, as far as possible, systematic and explicit and which permits the identification of all the "significant" events* which express the implications and therefore permit a better evaluation of the probabilities. These methodologies employed are the cause-consequence analyses (FMEA, operability analysis) and the consequence-cause analyses, (fault tree method).

In this phase the systematic and powerful help of the computer is reserved for the deductive part of the analysis, to which specialists in logics, mathematics, and reliability pay most attention. On the contrary, minor attention seems to be dedicated to the inductive cause-consequence part of the analysis, considered possibly more empirical, more extraneous, or more artisan by mathematicians and reliability experts, but which is determinant regarding the result. Many useless mathematical subtleties can sometimes be seen adopted for rough models of reality and are sometimes even entirely wrong models only because the responsibility of modeling is considered of different specialization.

The third phase is the one in which, through the inductive method of cause-consequence analysis, the knowledge relative to the object of the risk analysis is made clear, collected, organized, integrated, and finally completed. It is the success of this activity, aimed at the explicitation of the knowledge, which determines the success of the analysis, guaranteeing representation and a maximum possible objectivity, and is not the research of numerical-mathematical sophistication tied to deceptive and unfounded precisions or much worse the research of the "true" probabilities able, as some believe, to give credibility to each model, even though superficial, stereotyped, or careless. It is therefore in the process of collection, interpretation, and verification of the internal congruence of knowledge of research not of what is only significant but of the whole possible that the most appropriate, efficient, and powerful instruments must be used. If this happens one realizes that, after the inductive process of systematic collection and interpretation of knowledge and after the verification of its congruence, there is no longer any deductive process to carry out, because the logic of the system is completely clear.

It is sufficient at this point to attribute to each of the possible events identified its real weight, on a basis of the knowledge relevant to it, because the logical probabilistic model is completed and the different possible occurrences are clearly identified and quantified. Naturally, the aid of the computer is essential in this process and only its availability allows us to show clearly the process correctness and its characteristics of objectivity: in the complete transparency of the logic and in the absolute traceability of the knowledge used by the

---

* Pay attention to the conceptual difference between "significant", which is only what seems to be important to the analyst with a preliminary knowledge, and "possible" which seems important as a consequence of the analysis.

analyst. The characteristic formulation of the third phase, presented here, has its own conceptual validity which makes it independent from the moment, level, or grade of detail in which the analysis is to be made. It is always the knowledge we have that is explicit and from which the conclusions, formulated in terms of the uncertainties which it projects on the effective verification of undesired consequences, are taken. This is the Bayesian process, referred to previously. It is the only process which permits the continuous definition of the state of risk with reference to the preestablished risk objectives and so which lends itself as a basis for an effective quality assurance activity intended to satisfy such objectives.

In particular, during a design process, the analysis can take place at any moment and, at any moment, report nonconformities with the preestablished risk objectives requiring recourse to corrective measures such as different solutions at system, subsystem, and components levels, or as the searching of a greater knowledge through research or qualification tests. This characteristic is entirely peculiar to methodologies of analysis based essentially on inductive processes, while it is extraneous, except for evident strains and consequent solution complexities, to methodologies based on deductive processes.* In fact, to establish and demonstrate the tautologies proper to a not banal deductive process, it is necessary to have a solid and detailed knowledge of the particulars, which is not possible to obtain in the evolving phases of a design or realization.

Having established that the methodological approach to which we will refer in the following is that of logical probabilistic analysis of the inductive type, and in particular that effectuated using the dynamic characteristics of the event tree, we will now proceed with a succint demonstration of how it is possible to deduce progressively from general requisites those requisites of success assurance for the system functions and for the specifications of the components which assure them.

V. REQUISITES AND SPECIFICATIONS

Analysis, aimed at deriving the requisites of functions and, from these, the component specifications must be conducted on different levels from the general functional level to plant level, to system level, to subsystem level, and down to component level for all those components whose performance reliability seems critical. At plant level, it is necessary to determine and define the processes which occur, the states in which the lines of processes can evolve, and the actions which permit the transition from one state to the next. It is then necessary to subdivide the functions into functions necessary to the occurrence of the processes (or to the maintaining of the processes in balanced or pseudobalanced conditions), into protection-safety functions (which assure undesired transition to desired states), and into intervention functions (which allow the desired transition to desired states). The functions should be attributed to systems and distinguished in the actual functions of the system as they take place within the system itself, in functions borrowed from other systems, and in functions lent by the system to other systems. In this way, a functional plant model which allows the distribution among the functions of the requisites of success assurance necessary to satisfy the general objectives of risk established is realized. A knowledge of the results obtainable with the technologies available, not only in terms of performance but also in terms of their assurance within the desired margins of confidence, permits us on the one hand to put the critical choices into evidence and, on the other, to anticipate the necessities of research and development required to gain the lacking knowledge. Having made the choice and taken the decision on a general functional level, it is necessary to turn our attention, for each of the systems individuated, to the problem of their effective adequacy

---

* So tempting, accordingly with western culture which is attracted to the vision of the German idealism rather than to the observation of British empiricism (and the reason has been pointed out by Ernst Mach in *Mechanism and its Historical Critical Development*, in the conditioning imposed by the teaching-learning process).

in guaranteeing, with the desired margin, the success of the requested performances; those which at plant level have been modeled as modes of failure of the success conditions of the requested functions, now become failure modes in the systems due to malfunctioning of their subsystems or components.

Where possible, a unique model is elaborated, more and more detailed, which permits the insertion, in terms of probability, of the evaluation of margins of uncertainty relative to the nonverification of the failure modes of the individual functions, requested in the task conditions assigned or forecasted. The model obtained, in this way, simulates, according to a logical pattern, the behavior of the plant and provides, on an ever more-defined level, all of the events expected and the probability attributable to each one. The logical probabilistic model, where integrated with one or more phenomenological models, allows us then to obtain directly, in terms of probability and consequences, the evaluation of each forecastable event, furnishing the characterization in terms of risk of the effected choices, and permitting at each stage a direct comparison with the preassigned objectives of risk.

If the uncertainties in the definition of the task conditions or in the performance of components are such as to give place to an unacceptable evaluation of risk it is necessary to individuate different solutions to be adopted or the further knowledge to be acquired. In this way it is possible at each stage to evidence the corrective action to be taken, making simple the recourse to classical models of quality assurance.

The precept of correctly understood actions of quality assurance, aimed at acquiring knowledge of the characteristics which determine the behavior of components, subsystems, or systems, already constitutes a remarkable corrective action which requires to be quantified and consequently graduated, according to objective specifications, in order to obtain a maximum effectiveness.

It is the logical relationship between knowledge, uncertainty, and probability which permits the recognition of the role of technical specifications in the searching and that of the specifications of quality in the assurance (within the margins of confidence to be established) of the knowledge of the behavior of components (component in reliabilistic sense) and so in assigning them the characteristics of reliability to be inserted into the previsional models (rather than using average values of undifferentiated populations). Difficulty or impossibility to establish a single and omnicomprehensive model in the case of systems characterized by a too-high level of complexity has been mentioned above; this was intended by the words "where possible a single model is elaborated."

In these cases it is necessary to develop a more complex model which incorporates the results of the analyses conducted by means of a greater number of models, taking into account always more detailed levels. Whenever necessary, one should investigate the design specifications or the checking of such constituents, thus making the components lose the privileged role of supporting cornerstones of the reliability model, which a certain way of understanding the problems of reliability, based more on statistics than on engineering, would bring, without criticism, to them, also even in situations where the statistical aspect is quite doubtful if not totally uninfluential (components in small numbers or single components of specific plants).

In a particular or new situation it is necessary to know how to appreciate completely the purpose of every decision (including that of not deciding) and for this reason it becomes necessary to make recourse to risk analysis; the evaluations in such analysis are probabilistic, and not simple statistical projections, as sometimes and in some places should be desired, tending even to furnish true probabilities of single components. Conversely, with a single or new case, it is the analysis of the effective knowledge of the situation which alone permits the formulation of sensible forecasts which can be reevaluated on the basis of any further successive information.

The problem of a correct definition of the performances and of the assurance of their success (reliability) is, above all, a problem of engineering, in which statistics can be used as a support to engineering, but which cannot be in any way taken for a problem mainly, or even worse purely statistical.

It is possible, more so it is necessary, to work in terms of reliability, even in the absence of statistical data recordings, though precious, because each of us does not stop from making decisions only because of this absence, while, in deciding, it is necessary in every case to evaluate the risk connected to the state of knowledge. Only this evaluation could indicate the necessity of widening such a state, even through the recourse to statistical investigations of behavior, but well defined after an analysis in terms of engineering and not simply generic. Qualification and requalification therefore assume the role of essential cognitive steps with the aim of reducing risk, obtained through an increase in the basic engineering knowledge. It is such knowledge, in fact, which determines the margin of uncertainty about the avoidability of faults or accidents and which therefore alone legitimately allows a reduction in the evaluation of probability when it brings evidence of more guaranteed capacity.

There should be no fundamental disagreement or significant differentiation upon these concepts; when differentiations and difficulty begin, from concepts it is necessary to pass to operative methodologies, which permit the correlation of engineering knowledge, the uncertainties it involves, and the probability evaluations which result in an explicit unambiguous way with a rational process which is completely traceable and so univocally suitable to the actual state of knowledge, in accordance with the definition of objective method, which we want to reaffirm here, so leading to quantitative evaluations of probability which are also objective. Instruments for this are available; concrete proposals are not lacking.

VI. DESIGN AS AN ORDERLY DECISION-MAKING PROCESS

The realization path as a whole and each of its moments constitute activities which bring one to the identification of requests, starting from general performances (processes and functions which must take place and their quality) up to the requirements of final products. A list, although brief, of these moments must include conceptual design, preliminary design, overall system design, systems and components design, executive design, manufacturing of components, commissioning, mounting or assembly, systems testing, plant testing, operation, preventive and corrective maintenance, and modifications.

Essentially, a set of realization and operation specifications, initially synthetic and general (but not generic) must lead, through a series of logical and ordered steps, to more detailed stages, using successive decisional activities, up to the "frozen" stages of "as built", "as operated", and "as maintained".

In order not to become chaotic and not to run the risk of degenerating into an abnormal process which has lost sight of the objectives, the transition of the specifications to detailed levels which are more developed must, therefore, occur in a logical, ordered way and, last but not least, control the correspondence to the objectives. The development of the specifications and the verification of their correspondence to the objectives must, therefore, constitute a unique contextual process which is systematically memorized in both its essential components: "how" and "why". While proceeding with this development, it is important not to forget, and not to hide the fact, that the whole process constitutes substantially a succession of decisional acts. In fact, every major detail assumed, intrinsically, involves the introduction of some element of uncertainty, due to inevitable lacks in knowledge, and the presence of such uncertainty generates a necessity for decision. Step by step, therefore, the uncertainty inherent in the choices operated must be analyzed and quantified so that each decision is assumed in a conscious way which can be improved upon and not in an acritical and irreversible way. In particular, in the course of the decision-making process uncertainties must be confronted with objectives, preestablished on the basis of criteria of reasonable minimization of risk, whether of venture or of safety.

The control of completeness, suitability, and compliance of the specifications to the objectives, therefore, requests an on-line risk analysis which, besides the systematic and synthetic recording of the how, permits an explanation of the decision-making element of the why, and of its equally systematic and synthetic recording. In order to make this discussion clear, anchored to precise moments and facts, we can refer to the safety requirements of national and international regulations, in the field of realization of nuclear plants. Such regulations, in fact, expressely request that each of the fundamental decision-making stages of a planning process must be explained and documented in how and why, i.e.,

1.   The conclusive stage of conceptual design (preliminary design and related safety report).
2.   Conclusive stage of the overall plant design at systems level (overall system design and related safety demonstration of compliance with the safety goals of the preliminary design).
3.   Conclusive stage of design at system level of each relevant system constituting the plant (detailed system design and relative demonstration of compliance with the safety goals).
4.   Completion stage of each relevant system (reports on commissioning and demonstrating of compliance).
5.   Completion stage of the plant before the start of operation (report on testing and demonstration of compliance, operational technical specification, and related safety demonstrations of compliance).

The example does not stray from the field of conventional engineering but it serves because the information it requests constitutes a valid list of that which, although synthetic, is absolutely indispensable to the designer and to the operator in order to guarantee respect to the agreed safety goals. In fact, each of the above-mentioned moments constitutes the transition from one stage of development to a next, i.e., what in sporting language is called the moment of the passing of the baton from one runner to the next in a relay race. The handing over of a clear picture of how and why of the decisions which has been taken before guarantees that, during successive developments, it is possible to proceed in such a way as not to contradict the objectives sought before or in any way to displace them. It also allows us, in the case of nonconformances, to identify them at the right moment, starting the necessary corrective actions with effective promptness.

It is clear how these fundamental requirements do not change; when in certainly more up-to-date and substantially more correct optics, the goals are not placed only in terms of safety but extended from safety to reliability, operability, and maintainability, thus including the whole risk of the enterprise associated with the plant itself. With technologies which have demonstrated themselves to be hazardous, above all for the difficulties connected to their acceptance, it is a matter of assuming decisively the point of view of prevention rather than that which is obsolete because it is uneconomical, of correction *a posteriori*.

It is, therefore, a real process of assurance of reliability, availability, operability, maintainability, and safety which is made possible by the intelligent use of methodologies and techniques of risk analysis, meeting that need for total quality which should become more evident with technological progress and its level of sophistication, since the beginning of the planning phase. It is within this process, therefore, that risk analysis is placed as an essential tool in the most up-to-date design process.

VII. THE DIFFERENT PHASES OF RISK ANALYSIS METHODOLOGY

An extensive and systematic use of methodologies of risk analysis took place in the nuclear field when it was asked which were, in terms of a correct evaluation of risk, the results that had brought to the systematic adoption, sometimes acritical, of a regulation intent on prescribing and describing, rather than on fixing objectives and requesting the demonstration of their satisfaction: a how regulation without a rational why.

It is worthwhile to mention that regulations of this kind are often a consequence of aggressive industries intent on expansion into world markets; what major result can be reached in the search of new markets, than that of characterizing as safe its own product for how it is, omitting the why, and therefore suggesting all different products are unsafe, without appeal or possibility of comparison? The digression stops here however, without insisting on the damages which with time a similar apparent advantage causes by bringing to sclerotic state solutions which become technologically out of date. In the nuclear field, therefore, the application of the risk-analysis methodologies took place in the form of an effort of critical revision *a posteriori*, with all the unknown qualities and uncertainties which every revision *a posteriori* entails in the effective possibility of restoring a knowledge of the object under analysis, sufficient to model it in a correct way. In this situation the division among the phases of application of the methodologies has been pragmatic and due more to separation of different specializations than to logical internal necessity. The so-called probabilistic risk assessment (PRA) phases, although the probabilistic has already been discussed, relative to the plant (phase 1) to the containment systems (phase 2), and to external consequencees (phase 3), respectively, constitute a progressive amplification of the scope of analysis rather than an articulation in conceptual phases. The situation is completely different when the methodologies of risk analysis are framed according to easily identifiable phases of revision of design and realization stages, by a process of quality assurance, concerning only safety or also including reliability, availability, operability, and maintainability. This is in fact the tendency which can be seen emerging in the so-called conventional field, in those sectors in which the use of methodologies of risk analysis is already in existence or is already asserting itself. Therefore, there are, e.g., as far as the offshore drilling platforms are concerned, risk analyses which are expressely aimed at evaluation of the conceptual design, of the final design, of realization, and operation of the plant.

Further progress is taking place, and will be put into practice as soon as time allows a more diffuse assimilation by designers, using risk analysis as a design tool, thus conducting such analysis on line during design. As will be illustrated, today it is possible to model the plant in such a way that any modification or addition even of particulars can be inserted into the model in real time; the model can then be processed through computerized methods to generate the new risk scenario or scenarios on which to base the decisional process as in every other conventional method of design verification. The activity, specific to the moments of the freezing of the design or realization will become that of audit on the activities of risk analysis conducted and on risk evaluations, auditing aimed to ascertain the conformity to the requirements of correct application of the methodologies and the compliance of the decisions taken with the general preestablished goals of enterprise and safety.

VIII. GOALS AND CHOICE OF DESIGN

Speaking of goals and choices of design, requested for satisfying them, it is necessary above all to make the following basic statement: the true value of risk analysis lies in furnishing a valid instrument for the comparison of different situations, independent of the methodologies used, so long as the comparison is effected by a same methodology. Therefore comparisons of "absolute" results of risk obtained through different methodologies of risk analysis can and must be considered doubtful, while they are valid and significant if the same methodologies are employed. Therefore, the way of fixing the goals cannot be based on metaphysical considerations of the acceptable risk, but must be taken from careful analysis of the optimal possibilities offered by technology and from available resources. These analyses must be conducted using very precise methodologies which themselves become qualifying and essential elements of the goals as units of measure are, e.g., for measurements of length.

The results of these analyses consequently become the goals with which we must concern ourselves and in comparison with which verify the choices of design, realization, and operation, translating into practice general principles, which may be quite generic, such as that of "as safe as reasonably achievable".

In light of what was said above, in order to make the discussion about correspondence of objectives and fixed design criteria more realistic, it is necessary to make constant reference to very precise methodologies of risk analysis, conceived for and predisposed to the use in the design phase.

Here a further parenthesis is necessary: it concerns the intrinsic logic of methodologies of risk analysis. In general terms these methodologies are differentiated according to whether they are based on deductive or on inductive logic. A deductive logic is that which starts from a consequence and explores its possible causes in greater detail; a logic nearer to the inductive, instead starts from a cause and explores the possible consequences in temporal cascade. Examples of the two logics are the fault tree and the event tree, respectively.

According to whether a methodology of analysis is based prevalently on a deductive or on an inductive logic, it will be more suitable to the use in a revision process, or in a process of conception, which is typically that of design. In fact, the first one represents the instrument proper to the design testing laboratory or of troubleshooting, while the second is the logical instrument proper to the designer. It is this kind of opposition which brought people used to work with methodologies based principally on the fault tree to affirm that logical-probabilisitic analysis should be an instrument of verification which must refer to a fully realized object, not only to acquire its significance, but also simply so that it can be used. It is the same antithesis that has brought people who wanted to follow the evolution of a design from the beginning (contextually developing from this a representative and more detailed logical-probabilistic model which is, always, self-sufficient and able to permit verification of congruence with the preestablished objectives of each choice decided upon) to orientate themselves towards methodologies essentially based on event tree techniques: semiprobabilistic methodologies and, like their up-to-date evolution, probabilistic methodologies based on resolution of plant event trees.

IX. SEMIPROBABILISTIC METHODOLOGY

A. Objectives and Design Choices

Born in the nuclear field in the mid-1960s semiprobabilistic methodology was conceived to satisfy very precise basic requirements.

The first requirement was that of freeing the design intent on satisfying objectives of reliability, availability, and safety from the strict availability of statistical data. It has been necessary ever since then to emphasize how the research of reliability and availability data is not so much conditioned by statistical treatment of data in itself, but it is tied to the guarantees of behavior of the subsystem or component, which derive from the knowledge of its behavior (and this knowledge can well be drawn from experience of extensive use, but not exclusively).

It was necessary that every other method of acquiring or estimating knowledge able to furnish adequate guarantees should be considered usable as the recourse to historical data, for which it has then been intended to emphasize the problem of the criterion of being representative. In fields where the available and estimable knowledge is inadequate, it is wise to proceed by means of experimentation, either that more specifically related to reliability or that intended to deepen or clarify the fundamental knowledge. For this reason the equivalence "probability of success towards quality assurance and guarantee of functioning" has been set.

This treatment is coherent with the most modern interpretation of the significance of probability and therefore with the theorizing which derives from it.

Different from what we can believe at first sight, it is not therefore because of the characteristics of discrete subdivision of the range of probability and the assignment of intervals thus obtained into preestablished classes of guarantee that the name of semiprobabilistic has been assigned to the methodology; this name is derived instead from the necessity of satisfying a second requisite, with the methods available at that time.

This second requisite was that of rendering the methodology a work tool available to every single component of a design group, in every moment of its activity. It is in fact a basic criterion of quality assurance that everyone should have the knowledge and the instruments necessary to verify, by himself, the correspondence of his own work to the objectives which he has been given, and to find and promote personally the necessary corrective actions in the case of noticed nonconformity: "prevention is better than cure" *(melius prevenire quam reprimere)*. The compliance with this second requisite was very complex in that it was obtained by framing single branches of the event tree (single histories concerning events which took place according to defined modalities as far as their natural conclusion or the definitive block by the insertion of protective functions such as those which at a given moment are in front of the designer of a single part who is obliged to effect his choices on a basis of the only paths which are critical for him, into a complete vision of the entire plant situation, i.e., that which can be obtained only from the completed plant.

To say in a more concrete way, it is necessary that the individual designer, while making sure that each sequence he examines is within a field consequence probability, considered acceptable, reasonably guarantees also the automatic respect to the consequence-probability objectives for the whole plant sequences. For this purpose semiprobabilistic methodology operates in a triple way:

1. It defines the significance of sequence, prescribing that we do not get down to particularization of more details than the functional ones.
2. It establishes general rules of redundancy, independence, and diversification, in such a way as to guarantee that the common causes are not characterized by values of probability superior to the forecasted low values and further prescribes the carrying out of evaluation of the effects of redundancy taking into account a general criterion of common cause acting in a casual way.
3. It defines, conventionally on a basis of results of analysis effected at the foundation of the methodology, the maximum number of critical sequences, potentially critical from the point of view of the consequences, which can be forecasted to take place.

In this way it is possible to divide the plant risk equally among each of the critical sequences, taking note of the maximum total number forecasted, introducing also, as greater guarantee of the success of the final verification, adequate safety coefficients.

The third criterion is also automatically resulting in that the risk is equally divided among all the sequences which bring about comparable consequences.

Naturally, what has been adopted and described is not a quantitative process, but rather a semiquantitative one, which bring us to the desired results, a part verification to be effected *a posteriori*. This constitutes an empiric methodological passage, which has pushed us to call the entire methodology semiprobabilistic.

As will be seen in the following, the way of satisfying the second requisite has now substantially changed with the advent of a large scale of the computer, able to memorize all the data and design information and to make them available in real time to each component of the design group itself. Following this advent and these new possibilities, semiprobabilisitc methodology therefore evolved into probabilistic methodology based on the dynamic resolution of plant event trees, description of which will be given after the following brief illustration of the original semiprobabilistic approach.

B. Logical Bases and Fundamental Definitions

In the following conceptual assumptions, basic logic, and fundamental definitions of the semiprobabilistic methodology are given.

Conceptual assumptions may be summarized as follows:

1. No preestablished set of principles, criteria, standards, and codes, however large, can on its own meet initial requirements and satisfy safety aims in relation to time.
2. Initial requirements and the attainment of safety aims in relation to time can only be guaranteed if constant checks are carried out and all corrective measures required are suitably implemented. Therefore, aims must be clearly defined and be verifiable and checks must cover the whole life of the plant, from the design to the operating phase.
3. In order to be clearly defined and verifiable, aims must be quantified using measurement methods and techniques. Quantified aims, together with measurement and checking methods and techniques, constitute a nondissassociable and necessarily self-consistent unit (every quantified aim is necessarily associated with an objective definition of the method used to determine and check it).
4. The quantification of safety aims (aims at reducing risks) involves the simultaneous definition of (1) consequence of event considered and (2) margin of uncertainty (assessed in terms of probability) associated with the "ability" to deal (by means of preventive or protection measures) with the event considered.
5. Methods and techniques for measuring and checking safety aims must provide (1) physical assessment of the course of events considered and their consequences and (2) probabilistic assessment of margins of uncertainty of the "ability" to deal with events considered.
6. Consequences and associated probabilities must provide an estimate of the risk which must be optimized within the framework of operations carried out (in particular, the safety assessment aims at defining its "top end"). In no way can the scientific and purely metaphysical concept of "real risk" be allowed to intervene in this operation. Therefore, safety aims are set in terms of highest risk threshold whose nonattainment can be guaranteed in the light of information available and not in unproposable terms of "real risk" which cannot be known.
7. As a rule, the primary aim of safety activities is the "minimum reasonably obtainable risk" (obviously within the limits of the "proposable risk").
8. The analysis of what "is reasonably obtainable" within a practical context must allow for available resources of which cognitive ones are particularly important. Therefore, in the technological field, optimization means using proven technologies available which are applied on the basis of quality assurance procedures that can guarantee optimum results. It is a basic rule that, within the framework of solutions leading to similar results, the simplicity of the solution is a discriminating parameter.
9. Within a given technology, technical codes and official standards represent concise coding systems for proven solutions available and provide appropriate information as to their use which is confirmed by experience gained on the application. Therefore, it is essential to resort to them when determining the optimum application of proven technologies available.
10. Resorting to codes and standards does not however solve the problem of determining

the optimum application as this requires a constant check on completeness and consistency as well as careful analysis of the significance of the coded experience; this is an exhaustive risk analysis.

The need to adopt an analysis method which can be applied during every phase of the design, execution, and operation, and by every individual operator, and which thereby guarantees maximum efficiency when checking compliance with safety aims, has led to the decision of proceeding along individual accidental sequences thereby defining safety aims in terms of probability of events and associated maximum allowable consequences. In this way the criterion of consistency of the degree of protection (guarantee of nonoccurrence) has also been met with accidental sequences leading to similar consequences. Therefore, the application of the method involves a series of iterations which start on a general level to reach a detailed level (i.e., from the whole plant to individual systems and components) and which can, if necessary, go as far as the most basic parts.

Another important aspect is the functional role of systems, subsystems, or components which leads to an initial major functional subdivision between process functions (and associated systems) and insertion protection functions (and associated systems). In substance, this is the subdivision between continuous performance and availability to respond correctly to call.

The specification of the guarantee required from a process function to remain continuously within the field of success, for the whole period of assignment established, is therefore automatically translated into reliability requirements (or quality, if understood as guarantee) for the system, subsystem, or component performing the function. The margin of uncertainty of the "ability" of a function to guarantee success during the whole assignment can therefore be expressed in terms of relative rate of failure of the system, subsystem, or component performing it.

An insertion-protection function must obviously be capable of answering correctly a random call and the uncertainty is therefore translated in terms of relative unavailability, this being understood as the probability of the system, subsystem, or component performing the function not being capable of satisfying the call successfully.

The logic scheme into which the application of the semiprobabilistic method is subdivided can be summarized as follows:

1. Prior definition of a scale of consequences of incidental events vs. probabilities, to be assumed as project aims. This histogram of maximum allowable consequences vs. the interval of frequency of incidental events causing them can be deduced from the analysis of an optimized application of technologies and resources available.

2. For every process and insertion-protection system, identification of functions completed and analysis of physical events resulting from their nonoperation (event tree technique); depending on consequences, definition of reliability requirements of every function involved and, therefore, definition of reliability requirements to be met by the systems, translating them into quality requirements for the systems.

3. From quality requirements of systems, identification of reliability requirements, and therefore quality requirements of their components by means of analysis methods capable of establishing requirements needed for complying with aims of systems. These methods too have progressed with time from the initial fault tree methods to current complete and computerized event tree methods.

Systematic and subsequent measures which allow these logic steps can be further schematized as follows:

1. For every process and every process system of the plant (and its component) identi-
    fication of typical malfunctionings which can lead to incidental sequences, i.e., iden-
    tification of process function loss conditions.
2. Identification of insertion-protection functions (substituting or trip functions) antici-
    pated to prevent the continuation of incidental sequences or to minimize their con-
    sequences.
3. Development of event trees with particular emphasis on guarantees required from
    success conditions of insertion functions by developing at the same time the pheno-
    menology deriving from the various options.
4. Definition of failure rates, understood as success uncertainties relating to the assign-
    ment, for components causing malfunctionings initiating the accidental chain and
    correlation of residual margins of uncertainty and quality classes.
5. Definition of failure probability of insertion-protection functions, probability associated
    with both availability uncertainty, and uncertainty of the effective capacity of systems
    and components performing it, also in relation to uncertainties of physical conditions
    and, therefore, uncertainties of success definitions; definition of the correlation between
    margin of uncertainty and quality classes.
6. Checking of the consistency of quality requirements established for possible initiating
    causes and for insertion-protection measures, translated in terms of margins of un-
    certainty and therefore probability, with previously defined probabilistic goals.
7. Identification of "critical areas", i.e., those for which compliance with aims involves
    particularly stringent requirements in terms of success guarantees and therefore quality
    requirements to be met.
8. For systems or components providing functions included in critical areas, detailed
    analysis of the suitability of the design and analysis, experimentation, and testing
    specifications required for attaining quality levels or success confidence margins re-
    quired.
9. For the above systems or components, detailed analysis of testing, access, and main-
    tenance criteria required for meeting guarantees of correct operation.

The method, which we have only outlined, is as a whole particularly effective if applied
systematically and exhaustively during all phases of the design, the application for which
it was basically created. In this way, there is a gradual increase, in line with the progress
of project activities, from the level of detail in the definition of critical areas and therefore
level of depth of the analysis. In particular, it must be emphasized that a clear and basic
definition of critical areas and associated problems constitutes a primary requirement. In
fact, it would be inconceivable to carry out indiscriminately, for all components of an entire
plant, in-depth analysis and specification and execution control and quality assurance activities such as those generally required by points which can be associated with critical areas
and which are specific from the point of view of meeting safety goals.

The inability to identify these areas at a sufficiently detailed level so that they may be
dealt with through operating specifications both during the various phases of the design and
execution and later during operation, unavoidably leads to serious situations of nonconformity
with safety aims required. It is for this very reason that attention and priority efforts are
being transferred from development of increasingly sophisticated reliability mathematical
models to the engineering definition of system and component specifications and standards
which are both necessary and sufficient for guaranteeing services needed at the level of
confidence required. The use of simple algorithms and safety coefficients in the assessment
of the number of critical sequences has allowed complex situations to be managed even with
necessarily limited resources. The current development of the application of computers and
their potential have allowed many limitations, which were needed up to a few years ago,
to be overcome as explained in the next section.

Since it would be impossible to explain in detail the whole semiprobabilistic method, we shall limit ourselves to discussing one of the basic elements, in particular, that of defining margins of uncertainty and their translation into failure rates.

C. The Failure Rates

As already mentioned, the failure rate relating to a given and specific failure mode represents the margin of uncertainty of the capability of a component to bring to an end, immune from the fault considered, the assignment of unitary temporal duration assigned to it. This margin of uncertainty can naturally be assessed on the basis of different values of the relative confidence margin.

If the unitary assignment time is fixed at a yearly operating cycle, rates are expressed in events per year and have the meaning of the margin of uncertainty in relation to the success for such assignment.

If a complex plant is examined one finds classes of components with different grades of guarantee or margins of confidence in relation to the success of the assignment as mentioned above.

When operating on the basis of orders of magnitude, intervals of the margin of confidence (mc), and corresponding intervals of the margin of uncertainty (mu) can be identified as follows:

$$0 \quad > mc > 90\% \qquad 1 \quad < mu < 10^{-1}$$
$$90\% \quad > mc > 99\% \qquad 10^{-1} < mu < 10^{-2}$$
$$99\% \quad > mc > 99.9\% \qquad 10^{-2} < mu < 10^{-3}$$
$$99.9\% > mc > 99.99\% \qquad 10^{-3} < mu < 10^{-4}$$

The following failure rate intervals are associated with the above margin of uncertainty intervals:

$$1 \quad < mu < 10^{-1} \qquad 10^{0} \quad > r \text{ (events per year)} > 10^{-1}$$
$$10^{-1} < mu < 10^{-2} \qquad 10^{-1} > r \text{ (events per year)} > 10^{-2}$$
$$10^{-2} < mu < 10^{-3} \qquad 10^{-2} > r \text{ (events per year)} > 10^{-3}$$
$$10^{-3} < mu < 10^{-4} \qquad 10^{-3} > r \text{ (events per year)} > 10^{-4}$$

The following values are given to intervals of rates (useful for standardizing subjective opinions):

- Not improbable: $10^{-1} < r < 10^{0}$ (events per year); L_0
- Improbable: $10^{-2} < r < 10^{-1}$ (events per year); L_1
- Very improbable: $10^{-3} < r < 10^{-2}$ (events per year); L_2
- Highly improbable: $10^{-4} < r < 10^{-3}$ (events per year); L_3

The event resulting from the occurrence of several independent events becomes, if the events are

- Highly improbable: very improbable
- Very improbable: improbable
- Improbable: not improbable

This set of rules can be summarized in the table of correspondences (conventionally fixed):

Not improbable	1 (event per year)
Improbable	10^{-1} (events per year)
Very improbable	10^{-2} (events per year)
Highly improbable	10^{-3} (events per year)

In order to clarify the relationship between margin of confidence, subjective (standardized) opinions, and design, execution, control, and maintenance specifications, the following correspondences have been introduced:

1.   Level 0 (L0): normal types of failures of industrial components that are not or cannot be subjected to a specific control, inspection, testing, and maintenance system; normal or equivalent industrial quality level: 1 to 10^{-1} failures per year.
2.   Level 1 (L1): failures of components which, under operating conditions and regarding the function considered, meet requirements of a high industrial quality regarding type of design, construction, inspection, testing, and maintenance. For example, mention can be made of the quality level which can be attributed to systems under pressure and temperature (in relation to serious failures) following the adoption of codes such as ASME III, class 3: 10^{-1} to 10^{-2} failures per year.
3.   Level 2 (L2): failures of components which, under operating conditions and regarding the function considered, meet requirements of a very high industrial quality regarding type of design, construction, inspection, testing, and maintenance. For example, mention can be made of the quality level which can be attributed to systems under pressure and temperature (in relation to serious failures) following the adoption of codes such as ASME III, class 2. Minor failures of systems for which ASME III has been adopted must also be attributed to this level: 10^{-2} to 10^{-3} failures per year.
4.   Level 3 (L3): failures of components with high standards compared to those of the previous level because they undergo, due to their importance, more stringent design, construction, inspection, and testing than normally anticipated in industrial codes, including advanced ones (with the study of special conditions, possible adoption of higher safety margins, etc.) as well as preventive maintenance programs. For example, mention can be made of the quality level which can be attributed, for systems under pressure and temperature (in relation to serious failures), following the adoption of codes such as ASME III, class 1. Multiple components made up of a group of two or more individual components of a lower level, independent and separately controlled, are also assigned to level 3: 10^{-3} to 10^{-4} failures per year.
5.   Special level (LS), a special level which has also been defined: special conditions of component failure which occur at an almost negligible frequency. For example, if it has been proved and it is quite impossible that the particular failure condition will occur, except in the case of it being the outcome of failures of a lower level which can be detected sufficiently in advance to guarantee the possibility of safe intervention. In addition, multiple components made up of several individual components of level 3, independently and separately controlled, are also assigned to the level.

In fact, every component can break down in one or more ways and every failure mode is characterized by a given rate which, thanks to the identification between rate and margin of uncertainty in relation to the correct behavior, is closely correlated to that which goes under the definition of the quality level of the component in respect of the failure mode considered.

Checking of the quality level of a component in relation to a given failure mode is done using different techniques: from the use of direct experience to a subjective judgment up to the technique of the fault tree which allows a complete analysis of failure modes and definition of appropriate rates. In particular, the latter check is required for critical conditions, i.e., those requiring a high level of reliability, especially if no previous significant experience is available and the definition of an adequate experimental theoretical program is needed to confirm and support the adopted failure rates.

The scope of the analysis is a systematic investigation of causes that can lead to the failure mode considered and a definition of elementary failures and their combinations, which are important as far as the occurrence of the failure mode feared is concerned. Therefore, every failure mode can be associated with a specific rate using, depending on experience available and the greater or lesser criticality of the component:

1. Data based on experience, if existing experience is significant, where significant experience means data which are consistent with operating conditions anticipated for the component. (Failure data can be found in literature; for example, mention can be made of AECL-4607 *Reliability and Maintainability Manual — Process Systems*, 1974, WASH-1400 Reactor Safety Study 1975, bulletins of SRS, SRD, and UKAEA, etc.).
2. Previously given subjective classification by resorting also, for guidance purposes, to comparisons with known industrial standards mentioned above, if a reasonable subjective assessment is possible (if such an assessment is not critical and is only used to determine critical components).
3. Detailed analysis with the fault tree technique, as described below, when dealing with a critical or potentially critical component and experience available is not important. (This technique has been developed in accordance with ACL-4607 *Reliability and Maintainability Manual — Process Systems.*)

Every failure mode of a component is in fact the result of a weighted combination of failures of its basic parts. These combinations are determined using the fault tree. By using this technique it is possible to move from the failure mode to all groups of failure of basic parts which can cause it. The problem therefore relates to establishing rates or margins of uncertainty to be assigned to faults of basic parts. The basic parts of a component break down because of a variety of basic mechanisms which are typical of the various basic parts. These mechanisms operate differently depending on whether they relate to mechanical parts, electrical parts, electronic parts, etc. but can nevertheless be grouped into six typical categories. In fact, faults can be due to

1. Phenomena which lead to a fault under almost constant load conditions and within a short or long interval of time; they can be defined as stress mechanisms.
2. Phenomena which lead to a fault because of repeated or fluctuating stresses even if the stresses are below the failure value; they can be defined as fatigue mechanisms.
3. Phenomena which lead to a fault because of the sudden application of the load; they can be defined as impact mechanisms.
4. Phenomena which lead to a fault because of the deterioration of the service due to chemical or electrochemical action; they can be defined as corrosion mechanisms.
5. Phenomena which lead to a fault because of the deterioration of the service due to wear, consumption, aging, and prolonged exposure to radiation; they can be defined as wear mechanisms.
6. Phenomena which lead to a fault because of the deterioration of the service due to melting, vaporization, decomposition, sticking, and loss of characteristics because of thermal effects; they can be defined as temperature mechanisms.

Each of these failure mechanisms in turn depends on characteristic parameters or factors. For example, the first three mechanisms are basically determined by the extent and distribution of loads in relation to time and space, concentration of stresses due to form factors, defects, actual conditions of consistency, etc., and capability of materials to withstand stresses in relation to time and space. For mechanical parts they can be analyzed, for example, by means of fracture mechanics techniques.

Corrosion basically depends on chemical environment, electrical phenomena, thermal effects, presence of stresses, and physical-chemical characteristics of materials both external and internal.

Wear basically depends on friction phenomena, surface contact conditions, lubrification action, thermal dissipation phenomena, decomposition or recombination processes, precipitation or dissolution mechanisms, formation and migration of faults, etc.

Temperature basically depends on extent and distribution of heat generation in relation to time and space, thermal capacity characteristics, thermal transmission phenomena, and capacity and distribution of heat centers in relation to time and space.

In practice, for every basic part and every basic failure mechanism it is necessary to develop a specific fault tree whose input is an assessment of the reliability guarantee of the basic part in relation to the various factors which influence its failure rate (reliability guarantee is the margin of confidence; therefore, it is necessary to go back from the margins of confidence or better from the margins of uncertainty relating to the various factors, to the overall margin of uncertainty of the basic part in relation of the basic failure mechanism considered).

Every basic failure mechanism of a basic part is in fact influenced by a group of different factors with different guarantee levels depending on design methods and criteria, fabrication and control methods, storage and construction methods, operating conditions, and maintenance procedures. For each of those factors it is possible to assess a failure rate corresponding to the margin of uncertainty which must be associated with the capacity of the basic part in relation to the actual factor (e.g., design criteria, construction methods, and associated specifications).

The quantification of the probability of failure in relation to the guarantee that the factor is within the range of success (does not fail during the period of assignment) can be obtained on the basis of the following:

Guarantee	Failure rate per year
Very uncertain	10
More than uncertain on average	1
On average uncertain	10^{-1}
Moderately uncertain	10^{-2}
Not fully certain	10^{-3}
Practically certain	10^{-4}

Therefore, a fault tree is prepared for every basic failure mechanism and, through the combination of the various factors, it provides the failure rate of the basic part due to the basic failure mechanism. Every basic failure mechanism, together with its rate, contributes to the overall failure rate of the basic part through a weight factor which can be determined from the failure profile and which represents the very group of weights of the basic part and of individual basic failure mechanisms, determined on the basis of experience or by analogy (extrapolation). This provides a series of means which can lead to the failure rates of components, these being essential for analyzing margins of uncertainty of the functional capacity of systems. This also provides logical, although approximate and conventional, method to check the suitability of basic specifications (stress and environment) for design, design specifications (methods and references), fabrication and control specifications, warehouse and construction specifications, testing specifications, and periodic tests and maintenance for the purposes of guaranteeing a margin of uncertainty or failure rate which falls within aims established by guarantees required by the system.

The attribution of a quality level to the component based on the failure rate can, however, not derive directly from the rate itself but requires a further analysis of failure modes which lead to it. In fact, two situations can occur:

1.   The rate (margin of uncertainty) is mainly determined by failure rates of basic parts of the component and is therefore an intrinsic quality of the component.
2.   The rate (margin of uncertainty) is mainly determined by operations on the component which originate from different systems or components and, in this case, the component considered is not related to the actual cause of failure.

In the former case, the component is given the quality level corresponding to the frequency interval within which the failure rate falls. In the latter case, the analysis must be carried out further to determine real faults which cause the malfunction and to isolate most frequent failures in respect of which appropriate measures must be taken with specifications and quality assurance.

Regarding process systems, i.e., systems which must provide a continuous function to the process or processes in the plant to ensure that they remain within equilibrium conditions, guarantees are expressed in terms of maximum allowable frequency if the functions do not fall within the range of success, i.e., of the starting or sequences caused by situations of nonequilibrium of plant processes.

X. INFORMATIC MEANS AND EVOLUTION OF METHODOLOGY FROM SEMIPROBABILISTIC TO PROBABILISTIC

The above exposition of the semiprobabilistic methodology, though forcibly brief and consequently incomplete, cannot be concluded without a short description of its evolution due to the possibilities offered by the development in informatic technology.

The problem which the semiprobabilistic method tried to solve was substantially that of the thoroughness and systematicness of the analysis but the price for these was an apparent simplicity and the introduction of forcibly conservative safety factors, with a consequent loss of formal rigorousness. Such a price constituted, in fact, the condition of the involvement — in the verification of general goals — of every single design decision (design in a wide sense, to include in it realization and operation phases) and of every single designer, recalling him to the consciousness of his decisional role and to the relative impact on the total result, possibly in a direct and conscious way without the mediation of what could seem to him a mysterious ritual of initiates to reliability safety.

In the absence of a generalized recourse to informatic instrument, which was possible only later, a fragmentation of events was compulsory and had to be done according to previously established rules; successively their definitive integration *a posteriori* at the final verification should have been made.

On the contrary, a centralized memory is presently available, able to register in real time the single decisions, to arrange them, always in real time, in the whole design, to evaluate and indicate their congruence with the aims or their eventual lacks and the consequent necessity of recurring to corrective action. This allows each single event to be integrated into a whole and to be compared to the representation of such whole in the form of a dynamic and constantly updated pattern, not conditioned, therefore, by aprioristic and inevitably rigid schematizations. Obviously, the obtaining of such results on a practical level requires a careful application of rigorous methodologies, an availability of informative instruments and fitting algorithms, a conscious skillful and animated direction, and, from all operators, availability, openness, and, absolutely essential, intellectual honesty. All that should not be judged a utopia, otherwise every discourse on innovative technologies would become a utopia, too; the presupposition of their realization in facts and not only in words is the acquisition of a mentality and of an education peculiar to a team open to criticism and self-criticism, which are implicit to a confrontation with a reality in evolution, in that it tends to measure itself with the charm of progress; it is the spirit of Prometheus which must triumph, not pseudoacademic self-satisfaction, demagogic dishonesty, and corporate sectarianism.

The way thus opened and the available methodologies permit to solve the problems of CAD (computer-aided design) techniques, oriented not only towards hardware, but specifically to the solution of the problems of software in the design of complex systems.

A fundamental role in this field will be taken by all the forecasted events (or in a more diffuse diction "the bank of the forecasted events"); equipped with the instruments which permit its updating in real time, it will guide the "design" through a decisional development based on the whole set of the information available and, consequently, coherent to it and contextually documented. The chosen instrument for the generation and the updating of the whole set of forecasted histories of the plant events (history is a chronological sequence of events which is an event in itself) has been derived from the logic of the event tree, generalized to accept binary as well as multiple options (in order to represent relevant failure and operation modes). Such a computerized instrument furnishes complete and aseptic information, aseptic in that it is not influenced by the attitude of the analyst; the information may then be used differently, permitting a vision of reality — reality represented by the pattern that is both objective and referred each time to the point of view in question. Each time this point of view must concern the goals of safety, availability, maintenance, and operability, even in transient conditions, always taking into account the health risk, the enterprise risk, and the operation costs, according to a multiform and live vision of reality. The way of obtaining this will be outlined in the following paragraphs.

XI. EVENT TREE AS THE INSTRUMENT OF A COMPLETE PROBABILISTIC METHODOLOGY

The risk analysis sector aimed at the determination of possible events and at their probabilistic evaluation is here called logical probabilistic analysis and not reliability analysis, a term more common but perhaps less precise. The instrument proper to such analysis, in fact, must answer the double exigency of

- Verifying whether or not the event considered pertains to the possible events; if it is neither certain nor impossible, then it is the legitimate object of a probabilistic evaluation.
- Making such evaluation, taking into account the knowledge available and satisfying the principle of congruence within the judgment formulated.

The first of these exigencies has an exclusively logical character and is based on the logic of certainty; the second has the characteristic of an optimized judgment, which takes into account all the known circumstances and arranges them within the logic of probability. This duplicity of exigencies requires that the instruments of a correct analysis must present themselves as logical probabilistic instruments.

Available logical probabilstic instruments will not be listed here. It is necessary only to refer to the first paragraphs of this section, for an illustration of those generally known.

The popularity of each method has been largely the work of the demonstrated ability to confront ever more complex problems in a simple way. Clearly, those methods which better presented themselves to computerized application or for which the computerized solutions were individuated early, resulted favorably: However, it is also certain that the development of one method rather than another has been fundamentally conditioned by its intuitiveness, i.e., by its resulting from generalization of past experiences and therefore from immediate understanding. Dealing with inductive and deductive methods has certainly been important insofar as the opposition between inductive and deductive rationality, which has its roots in the distant past, which is probably peculiar to everyone and has been openly manifested since as far back as ancient Greek philosophy. The method most applied, most developed, and still widely in favor, as facts show, is the fault tree method, direct generalization, in our opinion, of troubleshooting techniques used in instrumentation, electrical, and electronic fields.

It is true that currents of opinion in favor of the event tree have always existed, but they have always needed to mark time when faced with a lack of convergence of efforts sufficient to furnish them with instruments adequate enough to remove them from a state of being, artisan methods, or suspiciously theoretical methods. The method of inductive reasoning of the event tree (and also of cause-consequence diagrams) apart from being the basis of the scientific method, has always been that of the designer who develops the concept and, while developing it, puts its solution to the test.

The deductive method of trouble shooting instead of the fault tree method is that which has characterized the activity of a posterior testing (e.g., that of design testing laboratories), a system testing which often means more correcting than understanding.

However it may be, by extension of use and for the number of computerized instruments developed for its application, the fault tree method still represents the principle method considered. It is true to say that, especially in the face of even more complex systems, the guarantee of correctness of application has given place to procedures which are, in their turn, even more complex; it is also true that the necessity of taking into account common causes, indirect conditioning, and effects of restorations into service, in a fundamentally static process, brings with it notable logical difficulties, which in their turn translate themselves into notable and not always justified mathematical difficulties; however, notwithstanding all this, the methods and codes in this field proliferate, both on the qualitative grounds of the determination of the cut sets and the quantitative grounds of the assessment of reliability and availability.

Methods based on the use of Markov matrices and those which make recourse to cause-consequence diagrams developed particularly in RISO from the work of Nielsen have also had notable application even though in more particular sectors; however, in all cases complex instruments for calculation result, which need long execution times or request recourse to simplification (subdivision into subsystems with low grades of interaction), which can be applied correctly only in specific cases. On the other hand, complexity is intrinsic to reliability calculations for complex systems, so much so as not to permit the use of fast running algorithms, as Rosenthal also noted in his communication to the Conference on Reliability and Fault Tree Analysis — UCLA — Berthley — 1974 entitled A Computer Scientifist Looks at Reliability Computations.

The ways of sensibly reaching the crux of these problems of complexity seem to be limited to two methods:

1. Divide the systems into subsystems with weak interactions.
2. Directly generate only the combinations of possible and significant events, not following all the theoretical combinations, only to later discard the impossible (illogical) or the insignificant (with a probability inferior to a reference basis).

This second approach is that which is followed in the event tree method of solution, spoken about here. Substantially, the basic consideration has been that an effective instrument of comparison of the logics of the design should be capable of proceeding to a systematic analysis which is both complete and coherent, following the inductive process of the designer through to its final consequences, through a successive series of "and if.....then...else..." which puts into light its aspects of phenomenological knowledge and, not less, of logical forecasting. For the reasons above mentioned the instrument must also be capable of generating all and only the possible events, excluding a priori the formulation of noncoherent options, and must reject all possible combinations of elementary events which are countersigned by probability values inferior to the prefixed base values and therefore can be assimilated as illogical options.

The setup code, which in the following will be named ARB, is able to generate and memorize the whole universe of events of interest, in the above-mentioned sense, furnishing the most exhaustive and impartial analysis of systems information (operation logic and conditional probabilities), which are fed to it as input. This information, on the other hand, constitutes the most complete synthesis of the available knowledge of the possible modes of behavior of the system in the various conditions in which it can find itself, a synthesis which in many aspects is comparable to the level of information furnished by the functional description of the system and of its complete FMEA. It is this knowledge which must be put to rigorous test to show its final consequences and, with them, put into light its eventual limitations, imprecisions, or omissions. This is the task which ARB performs; its output constitutes and documents the result of this analysis.

This output, furthermore, is often bulky (it must be taken into account that ARB can process up to thousands of options, generating also tens of thousands of options, generating also tens of thousands of complex events) and it is therefore necessary to have available a set of codes able to render accessible and manageable the great amount of information contained. The possibility of managing a mass of information, available in the form of histories of plant events and relative probabilities, in an intelligent way resulted in fact to be the greatest problem for the effective use of methods of the event tree type, a problem to which the solution has required experience of the application of the method and recourse to different instruments and approaches.

One of the first instruments deemed indispensable is a code capable of effecting a selective reading of the histories available, on a basis of logical research instructions, fed as input; this, hence named SEL, is particularly able to extract all those sequences of events which lead to an assigned top event.

A second instrument, named STOP, can then define which ones among these sequences represent minimal cut and can place them in a decreasing order of probability, and for each nonminimal cut it indicates the minimal cut from which it derives, allowing the identification of the possible presence of different top event, within the preassigned definition of top event. In fact, analysis of the most complex combinations not infrequently brings about the identification of those which are so differentiated, as forms or as consequences, that they must be considered on a phenomenological plane as different top events, rather than as different modes of a same top event. In general the presentation of potentially different top events requires ad hoc phenomenological analysis to cover the areas of cognitive uncertainties, which logical probabilistic analysis has demonstrated.

The examination of the cuts which lead to each top event enables us to show eventual incongruencies present in the information fed as input to ARB. Even more interesting in this sense is the information which is obtained from examination of the residual complex of histories, once all the sequences of events relative to a whole of exhaustive forecasted top events have been extracted by means of SEL. These complex events constitute, in fact, possible combinations of elementary events which are not forecasted either as presence or with better reason, as consequence; they show, therefore, a lack of combinatory imagination on the part of the analyst, or imperfections or omissions in the input data.

On the basis of these observations, it is now clearer how the process, through which we arrive at a universe of events in complete agreement with the knowledge and comprehension of the possible ways of functioning of the system, is an interactive process, made effective and rapid by the combined use of ARB and SEL.

This universe and more simply the input file from which it can be regenerated at any moment constitute at this point an exhaustive logical probabilistic analysis which is filed in an easily accessible form and can be used efficiently.

For each practical variation to be made to the system, a simple corresponding variation is made to the ARB input. The comparison of the previous universe of events with the up-to-date one allows the direct demonstration of all the consequences of the given modifications, each time deductible from the whole available knowledge of the system. Eventual gaps in information on new phenomenological courses are immediately shown and can be filled. The combined use of ARB, SEL, and STOP therefore permits the analysis of all top events which are of interest, the disposition of the whole set of the cuts relative to each top, the identification and putting into decreasing order of probability of all the minimal cuts, and the obtaining of the probability assessment of the top; it allows us to obtain the solution to the classical problems of probabilistic analysis in every complete, simple, and documentable way.

However, the information in terms of probability does not exhaust all the information that the set of developed algorithms gives us. For example, SEL can clearly furnish all the situations or histories in which the success of any of the functions of each system component is demanded. Eventually, because phenomenological analysis connects the histories of the system with the physical conditions which distinguish them, it is possible to obtain, on one hand, all those physical conditions into which the component must be able to intervene successfully, and, on the other hand, the definition of the success of the component in all forecasted conditions, and therefore the design criteria of its performances. It is therefore possible to weigh the design specifications of each component vs. the entire spectrum of conditions and performances which define its success and, furthermore, to ascertain that the successive variations in the system do not create more demanding conditions or do not require more restricting performances, once taken into account all the differentiated variations through the most systematic of analyses.

This last aspect, which connects the probabilistic logics of the system with the engineering specifications of its components, is fundamental in the design phase and constitutes one of the pillars of quality assurance in that, through a systematic, completely documented, and traceable approach, it minimizes the risk of being faced with a nonconformity in the final verification phase.

A further and fundamental step forward in the complete use of the availability information was made when the automatic integration was obtained of the results of the events analysis, furnished as a set of sequences of events (histories) and their relative probabilities, with the results of the phenomenological analyses being constituted by the consequences of the individual histories. Therefore, the concretization of the probability-consequence correlation is realized in a direct and personalized form for each of the sequences of possible events.

It is no longer necessary to group the sequences into families of consequences judged similar, or assign them to predetermined top events, to which emblematic consequences are attributed.

Every methodological heritage of the fault tree vanishes in this way, while the event methodology shows all its power, its logical rigor, and its formal elegance.

The results of synthesis which it furnishes are at once simple and complete:

1. The cumulative probability vs. entity of consequences function shows us the probability of not exceeding any preassigned consequence level.
2. The cumulative probability vs. risk function shows us the fraction of risk associated with any probabilistic cut.

3.   The risk distribution vs. consequence range function enables us to select the critical sequences from the point of view of the risk.
4.   The risk value of each individual sequence quantifies its contribution to the total risk.

Once the sequences significant in terms of risk are identified through analysis of the above-specified synthesis data it is possible to recall them from the whole of the events generated for an examination of the histories which they represent and to identify the events which they include, clearly reading both their logical course and the in-spacetime courses of the physically relevant parameters. In this way it is also possible to have, together with a synthetic view, detailed and aimed information to be used in each successive decision. In the case of modifications of the system being necessary, it would be possible to introduce them into the model in a simple and direct way, to be able then to proceed automatically to the complete assessment of the new risk conditions.

We know very well that to adequately illustrate advantages and problematics of a generalized utilization of the proposed methodology, a much deeper and more extensive demonstration would be necessary. In the practical impossibility of going beyond the brief outlines given, we will however try to make them more comprehensible by illustrating two examples of their use in the following paragraphs.

XII. FIRST EXAMPLE OF APPLICATION OF THE METHODOLOGY

The first example of application of the methodology is concerned with a classical probabilistic analysis. The plant object of the analysis is a gas/oil/water separator. First, it will be examined from a functional point of view.

The input of the plant is a mixture of oil, gas, and water, so the aim of the plant is to divide the three components and to give them in a separate stream as output, as schematically shown in Figure 1:

1.   Gas, partially used as fuel and the remainder sent to the torch.
2.   Water, sent to the water treatment system.
3.   Oil, sent to the storage tanks.

This is obtained through the coordinate intervention of the functions of several of the systems, which make up the plant.

The start of the analysis is the singling out of these functions, which entails the subdivision of the plant into separate and functionally well-defined systems. This method of going from generals to details can be applied again and again, taking the analysis from system level to subsystem level, component level, and even component elementary part level, if the plant complexity or the detail of the answers sought asks for this resolution level.

The resolution of a plant into more detailed physical and functional unities is essential to the orderly, comprehensive analysis of complex system, supported by documentary evidence; it is also correct as far as it does not bring prejudice to the necessary final syntheses.

The possibility of reassembling into a single mosaic the details of the separate analyses is acquired through the constant check of the physical and functional interfaces. The analysis of a system will not accordingly be directed only to the definition of its failure modes, as derived by the combinations of the failure modes of its components, but it shall also be directed to take the comprehensive recording of the state of the interfaces indeed that a system both receives information on the state of the functions and of the physical parameters, that the other functionally connected systems are supplying to it, and sends information to the other functionally connected systems on its own operational state. Figure 1 schematically represents the subdivision of the main functions of the plant, object of the present example of analysis, into systems.

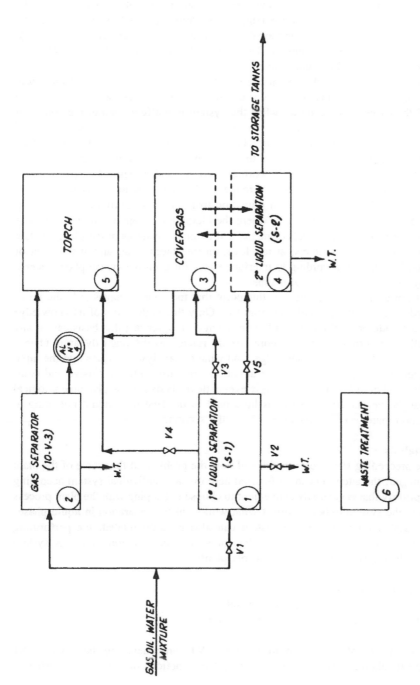

FIGURE 1. Schematic of gas, oil, and water separator.

For the sake of simplicity it will be assumed that the interfaces in input to the first of the systems be within the specified range, that is, flow and physical-chemical parameters of the mixture supplied through valve V1 meet the specifications of the plant. Valve V1 is the only interface to system 1 entrance; on the contrary, system 1 shows four interfaces at the exits that connect it physically and functionally to the water treatment system (valve 2), with the gaseous space of the flow-tank 10-S-1 (valve V3), with the torch (valve V4), and with the liquid space of the flow-tank (valve V5).

If the system is working properly the parameters at the outputs of the four interfaces will be within specification ranges and it will be considered as such in the analyses of the following systems. In all the cases to the contrary when the system is malfunctioning, the computer code bringing on the analysis memorizes these malfunctions both as failure modes of the system or as out of range of one or more interfaces. In this way both the functional inter-relationship between systems and the unity of the plant analysis will be met; the code (ARB) indeed automatically runs the interfaces, recording systematically their states (in range or out range, giving if necessary various out of range modes) and this in the phase of single system analysis as well as in the phases of transition from one system to the next.

How this happens is shown in Figure 2, where the hierarchic organization of the process of analyzing and recording at the levels of inputs and common systems of the plant, at state level of each of the common systems, at test level on interfaces state, and at the level of system analysis, brought out according to interface states for each one of the plant systems in a cyclic way, is outlined in a visual manner.

It is perhaps not useless to remember at this point that the code manages only the information received from the analyst with the input file. Only this is the basis of its knowledge and from this the code gets all the logically consistent consequences; this basis can in any way include all the information that is more strictly related to the plant dynamic phenomenology, what is in the analyst's knowledge, and what the analyst translates into the input file. It is deemed a main point to recall the attention explicitly to the exclusive and complementary relation that links together the logic eventistic analysis and the phenomenological analysis so that the deeper attention that is here paid to the first one should not bring to believe in a subordinate or even accessory role of the second.

A. System Analysis

Given in the preceeding short notes the idea of how the problem of the whole of the plant systems is managed, the attention can be focused here on each individual system according to an order of priority that is generally convenient to choose in keeping with the plant process flow. Therefore, the system to begin with is what works out the separation in liquid phase.

Now the requested functions for the system can also be individuated, the performing subsystems defined, the possible failure modes foreseen. The functions that this system performs are qualitatively the same for the whole plant:

- Separation of a gaseous phase
- Separation of a low density liquid phase, oil
- Separation of a high density liquid phase, water

The main constituents of the system are the valve V1 component and the separator S1 subsystem. The regulating functions and the protective functions of both can be listed as follows:

Valve V1
    Regulating function: fluid feed to separator
    Protective function: block on command

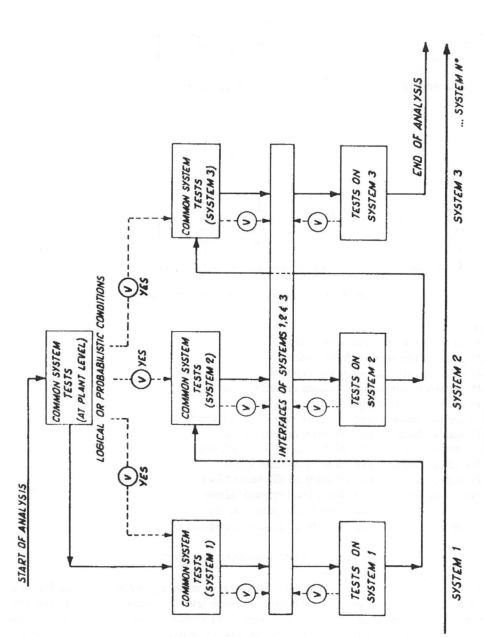

FIGURE 2. Logical procedure for functional analysis.

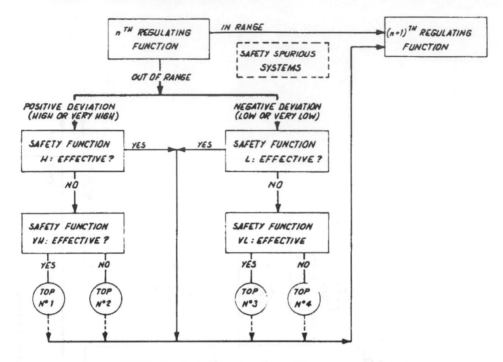

FIGURE 3.   Logical procedure for regulating system analysis.

Separator S1
Regulating function: pressure control
Regulating function: oil level control
Regulating function: water level control
Protective function: high pressure operator alarm
Protective function: high pressure bypass opening
Protective function: very high pressure V1 block command
Protective function: high oil level operator alarm
Protective function: very high oil level V1 block command
Protective function: low oil level operator alarm
Protective function: low oil level oil discharge block
Protective function: high water level operator alarm
Protective function: low water level operator alarm
Protective function: low water level water discharge block

The system is in this way divided into subsystems, each with its own regulating function and its own protective functions.

Now the dynamic event tree analysis methodology can be applied according to Figure 2 at plant level and to Figure 3 at regulating subsystem level in order to find out all possible operation modes and evaluate their probabilities. Each regulating subsystem is characterized by a limited number of operation modes (the operation modes are comprehensive in the fixed meaning of the failure modes too). These modes are chosen by the analyst in order to distinguish all the possible regulating system behaviors and to be able to continue the analysis leaving details behind.

In the following, these operation modes will be indicated for all three separator regulating functions; they will be given with their probabilities (calculated for 1000 h mission time) and with the number of the events, or different condition combinations, that have been associated to each operation mode through the selection operated by the code SEL on the events generated by the code ARB, with a probability cut off of 1 over 1 million (from about 100 to 200 events for each regulating function). Pressure regulation is characterized by pressure in range without protective intervention,

$$P = 0.798 \quad \text{number of events} = 1$$

pressure in range due to bypass intervention,

$$P = 6.04E\text{-}3 \quad \text{number of events} = 54$$

pressure under control due to V1 manual or automatic block,

$$P = 0.191 \quad \text{number of events} = 70$$

unprotected high pressure,

$$P = 2.15E\text{-}5 \quad \text{number of events} = 21$$

and unprotected low pressure

$$P = 1.11E\text{-}3 \quad \text{number of events} = 77$$
$$\text{Total} \quad P = 0.997 \quad \text{number of events} = 223$$

Oil level regulation is characterized by level in range without protective intervention,

$$P = 0.80 \quad \text{number of events} = 1$$

level under control due to V5 block,

$$P = 0.11 \quad \text{number of events} = 19$$

level under control due to V1 block,

$$P = 7.3E\text{-}2 \quad \text{number of events} = 65$$

very high oil level in S1,

$$P = 2.33E\text{-}4 \quad \text{number of events} = 39$$

and very low oil level in S1

$$P = 1.73E\text{-}2 \quad \text{number of events} = 25$$
$$\text{Total} \quad P = 0.997 \quad \text{number of events} = 148$$

Water level regulation is characterized by level in range without protective interventions,

$$P = 0.80 \quad \text{number of events} = 1$$

level under control due to V2 block,

$$P = 0.065 \quad \text{number of events} = 13$$

very high water level in S1,

$$P = 1.11E\text{-}4 \quad \text{number of events} = 13$$

and very low water level in S1

$$P = 0.13 \quad \text{number of events} = 75$$
$$\text{Total} \quad P = 0.997 \quad \text{number of events} = 102$$

Each failure mode (or better, each functioning mode) is in such a way characterized by a working definition and by a probability, giving the expected number of such events in the mission time. It is now possible on the basis of this information to use the system in a further modelization as if it were a single component.

XIII. SECOND EXAMPLE OF APPLICATION

An analysis, carried out as a demonstration of practicability in the course of an activity conducted on behalf of and under the supervision of ENEA-DISP, is illustrated as an example of application, in its entirety, of the developed probabilistic methodology.

The object of the analysis reported is a preliminary evaluation of the release of radioactive fission products, particularly cesium iodide (CSI) which could take place following a severe accident in a nuclear power reactor of the PWR type, with reference to the open literature data relative to the Surrey Plant (U.S.).

Normally, accidents which are defined severe are those which involve core melting and the loss of integrity of the containment but cases of undamaged containments will also be considered in the present analysis.

The dynamics of a severe accident can be considered as composed of three successive moments:

1. Events in the primary system which bring about core melt
2. Events which influence the behavior of the containment systems and which determine the released fraction
3. Dispersion of releases into the environment and their relative consequences

The purpose of this analysis is to answer exclusively the problematics relative to the second moment. Events, from the uncovering of the core to the release of fission products into the atmosphere, covering a maximum period of 24 h, will be taken into consideration here. As already mentioned, what will be reported here pertains only to the demonstration of practicability. It shall therefore be limited to the consideration of the case of core degradation and melting, caused by a complete loss of electric power. As we know, this eventuality is distinguished by its extremely low probability. However, in this analysis, which sets out by assuring the events as certain, it is by definition set equal to one.

As a consequence of the total loss of electric power, while conditions of unbalance between thermal energy from the core and thermal energy from the refrigeration of the plant remain, there is vaporization of the primary coolant, release of steam inside the containment through the safety relief valves, and the progressive reduction in the level of the coolant in the reactor vessel.

In what follows the successive evolution of events will be examined in detail from the uncovering of the core; the detailed analysis of the possible sequences of events, conducted within a conditional probability of 1 in 1 million, will enable us to identify as possible 3104 different incidental sequences.

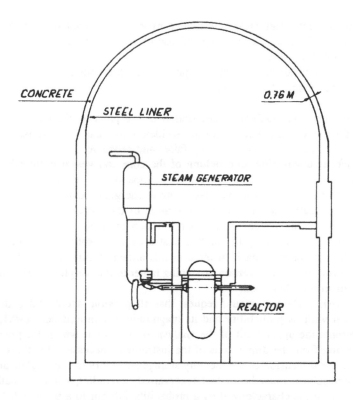

FIGURE 4.   Pressurized water reactor containment.

A. Description of the Dynamics of Accident

Figure 4 represents the outline of a PWR type reactor with the containment, including the primary circuit represented by the vessel, the heat exchanger and the pressurizer, and the containment spray system.

The dynamics of the accident is outlined according to the following phases:

- Initial event: total loss of electric power.
- Coolant boiling: decrease in the coolant level in the vessel followed by the release of steam from the safety valves.
- Overheating: uncovering of the core, Zr-H_2O reaction with production of heat and hydrogen, collapse of the clad with release of more volatile fission products, which in part are deposited on primary circuit surfaces which are colder, and in part are transported with the steam into the containment.
- Core melting: extensive phenomena of fragmentation and melting of the fuel resulting from uncovering of the central part of the core, great releases of volatile fission products follow. Steam explosions in the vessel, which could cause its collapse, are possible.
- Collapse of the vessel: in every case, as a result of the thermomechanic stresses consequent to the accumulation of molten material on the bottom head of the vessel, the bottom head collapse will follow.
- Drop into the reactor cavity: as a consequence of the vessel bottom head collapse, the molten material falls into the water in the reactor cavity. If this is in sufficient quantity a temporary solidification of the corium can be expected.
- Dispersion in the containment: the fall of molten material into the water can give place to steam-explosion reactions with further dispersion of corium, aerosol generation, effects on the temperature and pressure inside the containment.

- Corium-concrete interaction: the molten material can react with the concrete of the containment base with consequent release of aerosol and gases, including hydrogen and carbon monoxide.
- Containment breakage: the internal pressure on the containment can increase to the point of provoking its breakage.

Having presented the qualitative description of the physico-chemical conditions which can present themselves in the course of the accident under analysis, we pass to the division of the events into certainly true, certainly false, and probable.

Events such as uncovering and melting of the core and vessel bottom head failure are assumed certainly true and so are associated to probability 1; all other events are assumed to be only probable: from hydrogen deflagrations to energetic steam explosion, to consequent damage to the containment, and to the restoration of the electric power, with recovery of the containment cooling systems and airborne fission products removal.

On the basis of his knowledge the analyst must associate a probability value or a magnitude probability distribution with each of these events aimed at expressing his impossibility to exclude or affirm the event as certain, assigning rather to it a weight which defines his belief in its major or minor likelihood.

A general risk evaluation would request that all possible accidental paths be put into evidence, each with its probability and its magnitude of consequence. Obviously this objective represents the upper limit of the analysis only; a limit which for practical reasons; whether deriving from the dimensions of the instruments necessary for the successive synthesis or from the dimension of the memory storage necessary to accumulate all the accident sequences themselves, must be in fact reappraised, excluding from the paths taken into consideration all those characterized by a probability inferior to a preestablished threshold. In the example presented here the probability threshold is 1 in 1 million and so all accidental sequences having a probability conditional to core melt lower than that are excluded from the number of possible events. Of course the method of analysis applied automatically gives the value of the sum of the probabilities of all the sequences of events neglected in the result, and that to allow a prudent choice of the threshold value. In the present case, for example, it results that the excluded events amount to a total probability of 0.5 per 1000.

We refer to events in general rather than to the more specific accidental condition, in that the analysis conducted with the ARB code guarantees the generation of all sequences of events congruent to the functional description and data input and does not limit itself only to that of the accidental sequences.

It is evident, that as a guarantee of the exhaustion, in the definition of possible plant courses different approaches in simple situations are required (as compared with simple functional descriptions). In the first case, professional experience and the ability of single or small technical units can be sufficient; in the second, however, the availability of instruments able to cope with the elevated amount of knowledge initially available and that even higher amount of knowledge resulting from its deeper analysis is necessary.

To this end a chain of codes has been prepared which we refer to have and which has been used in the present example, illustrated in Figure 5. Its use begins with the schematization of the problem into a model on the basis of which to simulate, describe, and memorize all the possible modes of functioning or malfunctioning of the system under examination.

In the example illustrated here it seemed that a correct model of the dynamics of the events would need consideration of the phenomena of

- Deflagration of hydrogen, in the various phases of the accidental course.
- Steam explosion in vessel and reactor cavity. Direct heating, in concurrence with the release of molten material, at the amount of collapse of the lower vessel head.

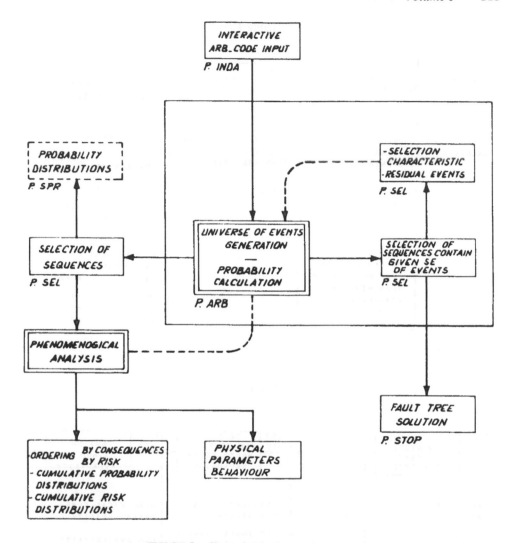

FIGURE 5. Chain of codes for decision tree analysis.

- The consideration of the function of electric power supply of which it is necessary to consider the possibility of restoration.
- The containment subsystem, depending on its functioning modes.
- Correct functioning.
- Lack of correct isolation in the initial phase of the accident (the releases in this case are indicated as due to preexisting openings.
- Collapse within the first 4 h (early openings).
- Collapse at various moments between 4 and 24 h (late openings).
- The subsystem of containment spray, refrigeration, and airborne removal of the containment of which restoration is considered possible on the return of electric power, if effectively operable at the moment of accident and not previously damaged by energetic phenomena (deflagration, steam explosions). Even after their restoration there remains the possibility of their being damaged by energetic phenomena taking place later.
- The component vessel of which it is necessary to consider the possibility of anticipated rupture from steam explosion.

Table 1

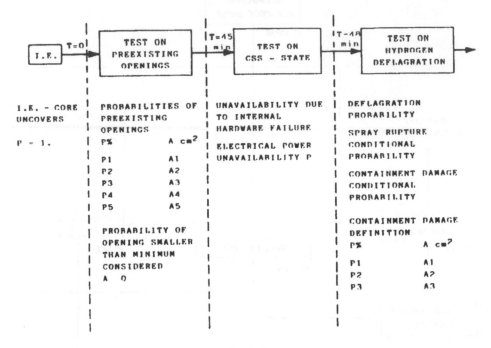

Table 2

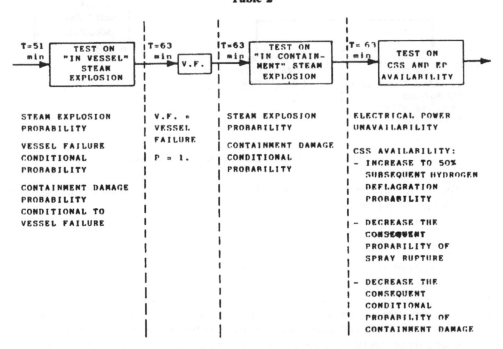

In Tables 1 to 4 we have an outline which synthesizes the adopted model of accidental conditions. Here they are reported, according to a temporal succession made from literature, alternatives attributed to events relative to phenomena, functions and behavior of subsystems or components, in unison with probability values assigned to them and with the conditionings they produce (probabilities that are conditional to them).

Table 3

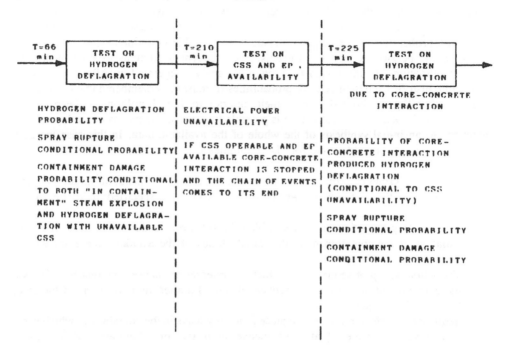

Table 4

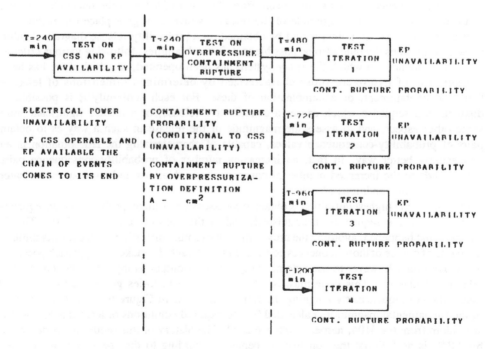

This description is transferred by the analyst into the ARB input file, in which the elementary events and their logical concatenation taken into consideration become even more precise and concrete, and construct a set of information sufficient enough for ARB to assess within the probability of preassigned threshold the possible combinations of events or, if preferred, the possible sequences of events. Such sequences amounts to 3104 in the case under examination. Having obtained, through the analysis operated by ARB, the universe of possible histories and their relative probabilities (it must be remembered that each history with a probability conditional to core melt, inferior to 1 in 1 million has been defined impossible and ARB treats it as such) it is possible to proceed, by means of the logical SEL selection, to an initial synthesis of the whole of the available data. In this case the 3104 sequences may be subdivided into four large families using the following criteria:

1.   The whole group of sequences in which there has been no isolation failure or casual breakage of the containment, which behave therefore as specified; there are 242 sequences.
2.   The whole group of sequences in which the releases are due to the lack of correct isolation of the containment in the initial phases of the accident; there are 996 sequences.
3.   The whole group of sequences in which the releases are caused by long-term leaks in the containment (in the interval between 4 and 24 h after the uncovering of the core); there are 768 sequences.
4.   Sequences in which releases take place following leaks in the containment which verify themselves within the 4 h after the uncovering of the core; there are 1098 sequences.

These four families could be directly used for a preliminary risk evaluation selecting from each one a sequence of events, significant from the point of view of the phenomena which characterize it and so of magnitude of the releases which they give place to; in this way four probability release points would be obtained on a probability-release diagram. Within these four families it would also be possible to define subfamilies, following various criteria, e.g., all sequences in which the spray system has been operated, in which there has been deflagration of hydrogen, those characterized by determined dimensions of leakage from the containment, or a combination of these. For each subfamily it is possible to distinguish a sequence which causes a conservative release, or a release significant from some other aspect, and to assess the magnitude of the release in such a way as to obtain pairs of probability-consequence values, capable of characterizing the risk situation in an increasingly better way. In fact, an increasing number of probability-consequence pairs corresponds to the increased number of consequence evaluations and so to an ever clearer definition of the conditions of risk.

Of course, the highest definition is obtained by assessing as many probability-consequence pairs as there are sequences distinguished, and in the present case this is 3104. This is exactly what happened, avoiding the provocation of eventual distortions in the results through the mode of choice of those sequences considered significant. To make this approach possible, it was necessary however to use automatically the ARB output as input for the code for the release calculations. This code must be able to read the histories generated by ARB and assess the consequences by adopting the right conditions. In Figure 6 we have the pattern of the fundamental magnitudes calculated for the required conditions of a sequence, selected at random from the 3104, namely sequence 2664. The history of this sequence, as described by ARB, is as follows: the containment behaves according to the specifications until the moment of partial collapse of its tightness, provoked by violent direct heating; at the moment of collapse of the vessel there is the deflagration of hydrogen released from the vessel with consequent effects of resuspension; and the electric power is restored 3.5 h after the un-

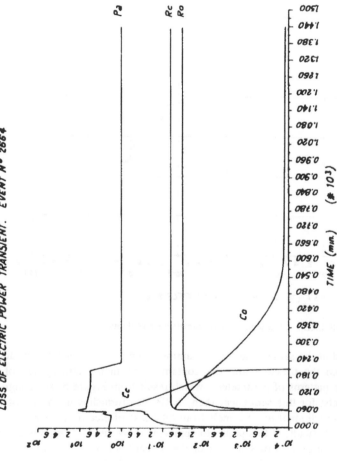

FIGURE 6. Sequence 2664 24-h histories of pressure, concentration, and release.

Cc · IN CONTAINMENT CONCENTRATION (10^{-4} m % OF INVENTORY/m^3) Pa : IN CONTAINMENT PRESSURE (ATM)

Co · IN AUXILIARY BUILDINGS CONCENTRATION (10^{-4} m % OF INVENTORY/m^3) Rc : RELEASE FROM CONTAINMENT (% OF INVENTORY)

          Ro : RELEASE TO THE ENVIRONMENT (% OF INVENTORY)

Table 5

Accidental condition	Sequence number	Probability (%)	CSI (% inventory, median release)
TMLB total of sequences	3104	99.95	7.9 E-3
Normal behavior	242	29.48	4.8 E-4
Failure to isolate	996	68.81	5.0 E-3
Late openings	768	0.73	2.3 E-4
Early openings (into 4 h)	1098	0.93	2.2 E-3

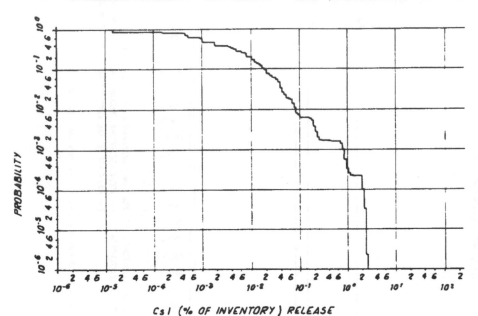

FIGURE 7. Consequences vs. cumulative probability.

covering of the core and at the same time the containment spray comes into action and in a short time lowers both the pressure and the contamination in the atmosphere of the containment. In fact, the patterns of parameters, which result from Figure 6, are congruent with this course of events; for this sequence, then, ARB also furnishes the value of 3.6/ 10,000 as a probability conditional to core melt from total loss of electric power.

It should no longer be difficult, at this point, to appreciate how much information can be obtained through the use of criteria of selection of the sequences combined with phenomenological analysis, which automatically assess their courses and consequences.

It is now necessary to present the last syntheses which may be derived from the analysis as a whole. They are substantially exemplified in Table 5 and Figures 7 and 8 and will now be briefly illustrated. Table 2 synthesizes the subdivision of the 3104 sequences in the four families just described. The accidental conditions which each represents, the number of sequences they comprise, the total probability of the family, and the release of CsI during the 24 h, averaged over the family and expressed in percent of the core inventory to represent the mean consequence, are defined.

Figure 7 reports on the x-axis the value, in percent of the core inventory, of the release of CsI during the 24 h to represent the consequences; on the y-axis the cumulative value of the probability, or the probability that the consequence is not inferior to the value in the x-axis, is reported. This plot demonstrates itself to be very useful in the decisional stage.

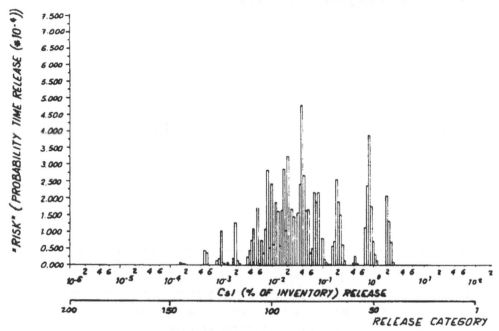

FIGURE 8. Risk distribution.

Figure 8 reports, in x-axis, the releases of CsI during the 24 h, always as a percentage of the core inventory and subdivided into 200 release classes. Substantially the 8 decades of the logarithmic scale which goes from 100 to 1/1,000,000%, have been subdivided into 200 equal intervals; all the sequences which have inclusive release within the extremes of each individual interval are assigned to the interval itself. The probability therefore associated to the interval is given by the sum of the probabilities of the sequences assigned to it, as the risk associated to it is given by the sum of the risk of the sequences assigned to it. This risk value is the magnitude reported on the y-axis. It also represents the contribution to an average release, or average consequence, of the class of the release, or consequence, individuated. Therefore, investigating the risk configuration, that is, which release classes, and within these, which sequences give significant contributions to the total risk, is immediate with the use of this graph. The structure of the subdivision into always equal release classes then permits a direct comparison of different accidental situations, providing them with risk distributions in homogenous form.

It is not thought either useful or possible to proceed with the illustration of the way in which the use of instruments of synthesis, as those mentioned here, permits rapid and effective conclusions, assessing and keeping in account all the information generated from the analysis.

Before closing, however, it is necessary to demonstrate how, by means of advanced methodology, it is easy to obtain sensitivity analysis in a form already suitable to decisional necessities. The problem for which we would like to present the solution is that of knowing how the panorama of risk varies if electric power is hypothetically not recoverable, instead of allowing for recuperability with increasing probability in time and equal, in fact, to 95% after 3 h, as in the above-reported analysis. To this end the ARB input is varied, feeding the information that the spray systems (the only ones to feel the effect of the restoration of the electric power) are permanently out of order. From this moment on the whole analysis proceeds in a completely automatic way and ends, on the part of the calculator, with the production of the plot in Figure 9.

CUMULATIVE PROBABILITY DISTRIBUTION - TMLB - $\left[\begin{array}{l} \text{WITH ELECTRICAL POWER RECOVERY (1)} \\ \text{WITHOUT ELECTRICAL POWER RECOVERY (2)} \end{array}\right.$

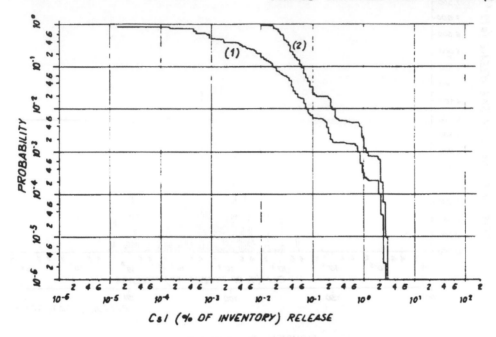

FIGURE 9.   Comparison between cumulative probabilities.

Here the two cumulatives are compared and it is easy to see how, equal being the probability, the consequences in the case of nonconsidered restoration are greater. Analogous and more circumstantial information can be deduced from the diagram in Figure 10 which represents the risk distribution.

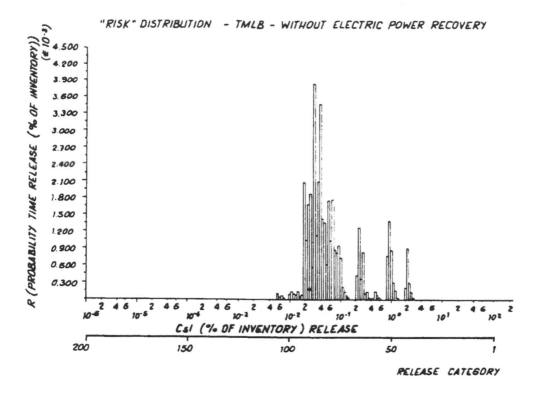

FIGURE 10.   Risk distribution — no electric power restoration.

REFERENCES

1. **Ragusa, S.,** Introduzione all'Analisi del Rischio nell'Industria, Safety Improvement, 1986.
2. **De Finetti, B.,** *Teoria delle Probabilita,* Vol. 1, 2, Einaudi, 1970.
3. **Farmer, F. R.,** *Reactor Safety and Siting: A Proposed Risk Criterion, Nuclear Safety,* Vol. 8, 1967, 539.
4. **Felicetti, F., Galvagni, R., and Zappellini, G.,** Analisi di Sicurezza — Metodologia Semiprobabilistica e suo Sviluppo Applicativo, CHEN-RT/DISP, No. 78, 1978, 10.
5. **USNRC, WASH 1400,** Reactor Safety Study, 1975.
6. **USNRC,** *PRA Procedures Guide,* Vol. 1, 2, NUREG/CR-2300, 1982.
7. **Swain, A. D. and Guttmanns, H. E.,** *Handbook of Human Reliability Analysis with Emphasis on NPP Applications,* NUREG/CR-1278, 1983.
8. **Bell, B. J. and Swain, A. D.,** A Procedure for Conducting a Human Reliability Analysis for NPP, NUREG/CR-2254, 1983.
9. **Nielsen, D. S.,** Use of cause-consequences charts in practical systems analysis, in Proc. Conf. Reliability and Fault Tree Analysis, University of California, Berkeley, 1974.
10. **Taylor, J. R. and Hollo, E.,** Experience with algorithms for automatic failure analysis, in Int. Conf. Nuclear Systems Reliability Engineering and Risk Assessment, Gatlinburg, TN, 1977.
11. **Melvin, J. G. and Maxwen, R. B.,** Reliability and Maintainability Manual — Process Systems, AECL-4607, 1974.

Chapter 5

SYSTEM RELIABILITY ASSESSMENT VIA SENSITIVITY ANALYSIS

Augusto Gandini

TABLE OF CONTENTS

I. INTRODUCTION

In the probabilistic analysis of systems, by which assessing the degree of confidence, under the point of view of performance and safety, which we may associate to a given system during its operation, it is clear that, beside the correct knowledge of the data base used, a key point is the identification of the important components. Such identification is in fact required in assisting the analyst in finding weaknesses in design and operation and in suggesting optimal modifications for system upgrade.

A number of methods have been suggested so far to evaluate the importance of basic components or of component-related entities, such as cut sets. We shall review here some of these methods. A new sensitivity methodology, mainly concerned with basic component parameters, is then proposed in relation to the Markov chain model and based on importance concepts so far widely used in nuclear reactor physics.[6]

II. CRITICAL BASIC EVENTS

Given the complemented system structure function $\overline{\phi}(\mathbf{X})$ (see Chapter 3, Section III) known also as top event, $\mathbf{X}$ representing the complemented elementary events $\overline{X}_n$ (n = 1,2, . . . N), N being the number of the system components, let us define functions

$$\overline{\Phi}(1_i,\mathbf{X}) = \overline{\Phi}(\overline{X}_1,\overline{X}_2,...\overline{X}_{i-1},1,\overline{X}_{i+1},...\overline{X}_N) \tag{1}$$

$$\overline{\Phi}(0_i,\mathbf{X}) = \overline{\Phi}(\overline{X}_1,\overline{X}_2,...\overline{X}_{i-1},0,\overline{X}_{i+1},...\overline{X}_N) \tag{2}$$

A basic complemented event $\overline{X}_i$ is said to be critical, for a given up (i.e., working) state of the system considered, if its occurrence (corresponding to the failure of the component related to it) will also correspond to the system failure, i.e., if

$$\overline{\Phi}(1_i,\mathbf{X}) - \overline{\Phi}(0_i,\mathbf{X}) = 1 \tag{3}$$

III. BIRNBAUM IMPORTANCE

Consider the expected top event $E\{\overline{\phi}(\mathbf{X})\}$ and the component expected event $E\{\overline{X}_n\}$ (n = 1,2 . . . N). We then denote

$$g_s(t) = E\{\overline{\Phi}(\mathbf{X})\} \tag{4}$$

It will be

$$g_s(t) = \begin{cases} \overline{A}_s(t) & \text{(system unavailability) if components} \\ & \quad \text{are repairable} \\ \\ \overline{R}_s(t) & \text{(system unreliability) if components} \\ & \quad \text{are not repairable} \end{cases} \tag{5}$$

Likewise we denote

$$q_n(t) \equiv E\{\overline{X}_n\} = \begin{cases} \overline{A}_n(t) & \text{(component unavailability) if} \\ & \text{components are repairable} \\ \\ \overline{R}_n(t) & \text{(component unreliability) if} \\ & \text{components are not repairable} \end{cases} \qquad (6)$$

Moreover, assuming components are independent, function $E\{\overline{\phi}(\mathbf{X})\}$ results in a linear function with respect to $E\{\overline{X}_n\}$, since $\overline{\phi}(\mathbf{X})$ is linear with respect to $\overline{X}_n$. We can then write, denoting by $q(t)$ the vector representing the quantities q_n,

$$g_s(t) = g_s[q(t)] \qquad (7)$$

If we expand function $g_s(t)$ around reference values $q_n(t)$ ($n = 1,2, \ldots N$), we can also write the expression

$$g_s(\mathbf{q}') = g_s(\mathbf{q}) + \sum_{n=1}^{N} \left.\frac{\partial g_s}{\partial q_n}\right|_q (q_n' - q_n) +$$

$$\frac{1}{2} \sum_{n,m=1}^{N} \left.\frac{\partial^2 g_s}{\partial q_n \partial q_m}\right|_q (q_n' - q_n)(q_m' - q_m) + \ldots \qquad (8)$$

The derivative $\partial g_s/\partial q_n$ represents the first order sensitivity of the system unavailability (unreliability) with respect to q_n. Due to its linearity of g_s with respect to q_n and considering that the range of q_n is $(0 \div 1)$, it can be expressed as

$$\frac{\partial g_s}{\partial q_n} = g_s(q_1,q_2,\ldots q_{n-1},1,q_{n+1},\ldots q_N) - g_s(q_1,q_2,\ldots q_{n-1},0,q_{n+1},\ldots q_N)$$

$$\equiv E\{\overline{\Phi}(1_n,\mathbf{X}) - \overline{\Phi}(0_n,\mathbf{X})\} \qquad (9)$$

This sensitivity coefficient is known as the Birnbaum importance[1] $[I_n^B(t)]$ and represents the probability that the nth component is critical. In fact, as we can see from Equation 8 and limiting consideration to first order effects, multiplying it by $\delta q_n(t)$, i.e., by an increment of the probability that the nth component has failed at t, gives the marginal increment of the probability that the system is failed by this same time.

IV. CRITICALITY IMPORTANCE

Let us divide Equation 8 by $g_s(t)$ as defined by Equation 5 and consider relative increments $(q_n' - q_n)/q_n = -1$ ($n = 1,2, \ldots N$). Since in this case $g_s(\mathbf{q}') = 0$, we obtain

$$0 = 1 - \sum_{n=1}^{N} \frac{\partial g_s}{\partial q_n} \frac{q_n}{g_s} + \frac{1}{2} \sum_{n,m=1}^{N} \frac{\partial^2 g_s}{\partial q_n \partial q_m} \frac{q_n q_m}{g_s} + \ldots \qquad (10)$$

If q_n are small values (as usually occurs) and we drop second order and higher order terms, we can then write

$$1 \approx \sum_{n=1}^{N} \left(\frac{\partial g_s}{\partial q_n} \frac{q_n}{g_s}\right) \qquad (11)$$

The term $[(\partial g_s/\partial q_n)(q_n/g_s)]$ ($n = 1,2, \ldots N$) is what Lambert calls criticality importance[2] $[I_n^{CR}(t)]$, corresponding to the conditional probability that the system is in a state at t such that the nth component is critical and is failed, given the system has failed this same time.

V. BARLOW-PROSCHAN IMPORTANCE

Let us now differentiate $g_s(t)$ with respect to time. We obtain

$$\frac{dg_s}{dt} = \sum_{n=1}^{N} \frac{\partial g_s}{\partial q_n} \frac{dq_n}{dt} \tag{12}$$

Considering the integral of this derivative over the period $(0 \div t_F)$, representing the mission time, we can then write the obvious relationship (identity)

$$1 = \sum_{n=1}^{N} \frac{\displaystyle\int_0^{t_F} \frac{\partial g_s}{\partial q_n} \frac{dq_n}{dt} \, dt}{\displaystyle\sum_{m=1}^{N} \int_0^{t_F} \frac{\partial g_s}{\partial q_m} \frac{dq_m}{dt} \, dt} \tag{13}$$

The term

$$\frac{\displaystyle\int_0^{t_F} \frac{\partial g_s}{\partial q_n} \frac{dq_n}{dt} \, dt}{\displaystyle\sum_{m=1}^{N} \int_0^{t_F} \frac{\partial g_s}{\partial q_m} \frac{dq_m}{dt} \, dt} \qquad (n = 1,2,...N) \tag{14}$$

is called the Barlow-Proschan importance[3] (I_n^{BP}) and corresponds to the conditional probability that the nth component causes the system to fail in the time interval $(0 \div t_F)$, given the system is failed in the same period.

VI. CUT SET IMPORTANCE

Importance expressions similar to those given above for system components can be obtained in relation to cut sets. In particular, we can define the Barlow-Proschan cut set importance (I_k^{CS}) (k = 1,2, . . . K), K being the number of minimal cut sets, as given by the ratio

$$I_k^{CS}(t) = \frac{\displaystyle\sum_{j=1}^{J_k} \int_0^{t_F} [g_s(1^{(k)},q) - g_s(0_j,1^{(k)-j},q)] \prod_{\substack{i=1 \\ (i \ne j)}}^{J_k} q_i \frac{dq_j}{dt} \, dt}{g_s(t)} \tag{15}$$

where $g_s(1^{(k)}, q)$ represents the value of g_s corresponding to the case in which all the J_k unavailabilities $q_{k,i}$ (i = 1,2,...J_k) of the J_k components relevant to the kth cut set are assumed equal to 1, whereas $g_s(0_j,1^{(k)-j},q)$ represents the value of g_s corresponding to the case in which all, but the jth one, unavailabilities of the J_k components relevant to that same kth cut set are assumed 1, while that relevant to the jth component is assumed 0.

Since to a cut set with all its elements equal to 1 (which is a condition equivalent to assuming equal to 1 the unavailability of all the components relevant to it) there corresponds a system unavailability g_s equal to unit, Equation 15 can then be written

$$I_k^{CS}(t) = \frac{\displaystyle\sum_{j=1}^{J_k} \int_0^{t_F} [1 - g_s(0_j,1^{(k)-j},q)] \prod_{\substack{i=1 \\ (i \ne j)}}^{J_k} q_i \frac{dq_j}{dt} \, dt}{g_s(t)} \tag{16}$$

VII. NONCOHERENT SYSTEMS

The above importances have been defined for coherent systems. For noncoherent ones it may occur that for some components the sensitivity $\partial g_s/\partial q_s$ is negative. This is natural since for noncoherent systems the complement of the complementary event $\bar{x}_n$ [$\equiv X_n$] may be critical. In these cases the sign may be retained as indicative of noncherence,[4] while the absolute value $|\partial g_s/\partial q_n| \equiv \partial g_s/\partial \bar{q}_n$, where $\bar{q}_n = 1 - q_n$, will again correspond to the probability of the nth component to be critical.[5]

VIII. RELIABILITY ANALYSIS VIA GPT METHODS

We have illustrated above the concept of importance relevant to components, or to component-related quantities, as cut sets. We shall introduce now a different, heuristically based, concept of importance, associated with a state, rather than with a component, within the frame of the Markov chain representation of the system state evolution (see Chapter 3, Section V).

Given a reference system formed by N components and to which K states are associated, recalling Equation 107 of Chapter 3 the state probability vector $\mathbf{P}(t) \equiv |p_1(t) \ p_2(t) \ \ldots \ P_K(t)|^T$ associated with such system will be governed by the equation

$$\frac{d\mathbf{P}}{dt} = \mathbf{M}_s \, \mathbf{P}(t) \tag{17}$$

with initial conditions

$$\mathbf{P}(0) = |1 \ 0 \ 0...0| \tag{18}$$

Let us then consider a quantity of interest associated to vector $\mathbf{P}(t)$, for example, the system unavailability averaged over the mission time t_F, defined as

$$\widetilde{\overline{A}}_s = \frac{1}{(t_F - t_0)} \int_0^{t_F} \mathbf{h}_0^{+T} \mathbf{P} \, dt \tag{19}$$

where $\mathbf{h}_0^{+T}$ is the transposed of vector

$$\mathbf{h}_0^+ = \begin{vmatrix} 0 \\ 0 \\ \cdot \\ \cdot \\ 0 \\ 1 \\ 1 \\ \cdot \\ \cdot \\ 1 \end{vmatrix} \begin{matrix} \text{line 1} \\ \text{line 2} \\ \cdot \\ \cdot \\ \text{line I} \\ \text{line I + 1} \\ \text{line I + 2} \\ \cdot \\ \cdot \\ \text{line K} \end{matrix} \left. \begin{matrix} \\ \\ \\ \\ \end{matrix} \right\} \text{operating (up) states} \qquad \left. \begin{matrix} \\ \\ \\ \\ \end{matrix} \right\} \text{failed (down) states} \tag{20}$$

Setting

$$\mathbf{h_a^+} = \frac{1}{(t_F - t_0)} \mathbf{h_0^+} \tag{21}$$

we can then write

$$\widetilde{\mathbf{A}}_\mathbf{a} \equiv \int_0^{t_F} \mathbf{h_a^{+T}} \, \mathbf{P} \, dt \tag{22}$$

In order to heuristically define the importance associated to a given state, say the kth one, at time t, let us assume that the probability $P_k(t)$ of the kth state at time t is increased by an amount* $\delta P_k(t)$. The average unavailability $\widetilde{\mathbf{A}}_\mathbf{a}$ will correspondingly change by an amount $\delta\widetilde{\mathbf{A}}_\mathbf{a}$. The importance, $P_k^*(t)$, associated with the kth state at t will be defined as the limit of $\delta\widetilde{\mathbf{A}}_\mathbf{a}/\delta P_k$ for $\delta P_k \to 0$.

The above importance definition implies the linearity of the Markov process, i.e., that the elements of the transition matrix $(\mathbf{M_a})$ are P-independent. Otherwise, for nonlinear systems, a marginal importance should be defined, corresponding to a system linearized around a reference solution.

Making use of the formulations relevant to the generalized perturbation theory (GPT),[6,7] we can write the general equation governing the importance vector function

$$\mathbf{P^*(t)} = \begin{vmatrix} P_1^*(t) \\ P_2^*(t) \\ \cdot \\ \cdot \\ \cdot \\ P_K^*(t) \end{vmatrix} \tag{23}$$

It results

$$-\frac{d\mathbf{P^*(t)}}{dt} = \mathbf{H_a^*} \, \mathbf{P^*(t)} + \mathbf{h_a^+} \tag{24}$$

where $\mathbf{H_a^*}$ is obtained by reversing operator

$$\mathbf{H_a} = \frac{\bar{\partial}(\mathbf{M_a \, P})}{\partial \mathbf{P}} \tag{25}$$

(for the definition of $\bar{\partial}(\mathbf{M_a P})/\partial \mathbf{P}$, corresponding to a Frechet derivative, see Appendix A).

Assuming Equation 17 is a system of linear equations, then $\mathbf{H^*} = \mathbf{M_a^T}$. Since it is known, by definition of importance, that $\mathbf{P} \, (t > t_F) = 0$, for the integration of Equation 24 it is convenient to start from t_F and proceed backward.

Consider now changes $\delta\alpha_i$ of parameters α_i (i = 1,2, . . .), representing component hazard rates, repair rates, etc. and then characterizing the system considered. The average

---

* It should be reminded that a change δP_k of the kth state probability, as induced by a system parameter alteration, is generally accompanied by changes δP_h (h $\neq$ k) of all the other state probabilities, such that $\sum_{k=1}^{K} \delta P_k = 1$. This of course does not detract from the general scope of the definition of importance given in the following, where only a single change δP_k is considered.

unavailability $\widetilde{A}_s$ considered above, will be changed by an amount $\delta\widetilde{A}_s$. Recalling Equation A.12 of Appendix A, it results (direct effects not existing in this case)

$$\delta\widetilde{A}_s = \sum_i \delta\alpha_i \int_0^{t_F} P^* \frac{\partial M_s}{\partial\alpha_i} P \, dt \tag{26}$$

The sensitivity, $\tilde{s}_{A,i}$, of $\widetilde{A}_s$ with respect to parameter α_i then results

$$\tilde{s}_{A,i} = \frac{\partial\widetilde{A}_s}{\partial\alpha_i} = \int_0^{t_F} P^* \frac{\partial M_s}{\partial\alpha_i} P \, dt \tag{27}$$

From the above equations we can see that perturbation $\delta\widetilde{A}_s$ and the sensitivity $\tilde{s}_{A,i}$ can be obtained by simple integration procedure in terms of functions P and P^* calculated at unperturbed conditions.

Another quantity of interest to system analysis is the system unreliability $\overline{R}_s(\tilde{t})$ at a given time $\tilde{t}$. In this case, the quantity of interest can be written as

$$\overline{R}_s(\tilde{t}) = |0 \ 0 \ ... \ 0 \ 1| \begin{vmatrix} \hat{P}_1(\tilde{t}) \\ \hat{P}_2(\tilde{t}) \\ \cdot \\ \cdot \\ \hat{P}_i(\tilde{t}) \\ \hat{P}_{abs}(\tilde{t}) \end{vmatrix} \tag{28}$$

where $\hat{P}_1(\tilde{t})$, $\hat{P}_1(\tilde{t})$, ... $\hat{P}_i(\tilde{t})$ correspond to operating (up) states, whereas $\hat{P}_{abs}(t)$ to the absorbing one. Setting

$$\hat{h}^{+T} = |0 \ 0 \ ... \ 0 \ 1| \, \delta(t - \tilde{t}) \tag{29}$$

$\delta(t - \tilde{t})$ being the Dirac's function, Equation 28 can be written

$$\overline{R}_s(\tilde{t}) = \int_0^{t_F} \hat{h}^{+T} \hat{P}(t) \, dt \tag{30}$$

Recalling Equations 117 and 115 of Chapter 3, vector function $\hat{P}(t)$ results governed by equation

$$\frac{d\hat{P}}{dt} = M_r \hat{P}(t) \tag{31}$$

whereas the corresponding importance function $\hat{P}^*(t)$ obeys equation

$$-\frac{d\hat{P}^*(t)}{dt} = H_r^* \hat{P}^*(t) + \hat{h}^{+T} \tag{32}$$

where H^*_r is the transposed of matrix $H_r \equiv \overline{\partial}(M_r P)/2P$. Vector h^+ in this case can be omitted from the above Equation 32, since it now merely corresponds to the final condition from which starting in the (backward) equation integration procedure. More exactly, the final condition is represented by vector $|00 \ldots 0\,1|$ appearing at the right hand side of Equation 29.

Analogous expressions of the perturbation $\delta\overline{R}_s(t)$ of the system unreliability $\overline{R}_s(t)$ and of its sensitivity, $s_{\overline{R},i}$, with respect to the generic parameter $\alpha_i(i = 1,2,\ldots)$ can be obtained, as those we have seen above for the system unavailability, i.e.,

$$\delta\overline{R}_s = \sum_i \delta\alpha_i \int_0^{t_F} \hat{P}^* \frac{\partial M_r}{\partial \alpha_i} \hat{P}\, dt \tag{33}$$

$$s_{R,i} = \frac{\partial\overline{R}_s}{\partial\alpha_i} = \int_0^{t_F} \hat{P}^* \frac{\partial M_r}{\partial\alpha_i} \hat{P}\, dt \tag{34}$$

The sensitivity coefficients obtainable with the GPT methodology can be directly related with the importance definitions given in previous sections. Consider, for example, the Birnbaum importance $I_n^B(\hat{t})$ at time $\hat{t}$ and relevant to the nth component, expressed by Equation 9. Introducing the (generally easily calculable) derivative of the nth component unreliability with respect to a characteristic parameter of such component, say the hazard rate λ_n, i.e., $\partial q_n/\partial\lambda_n$, we can generally write [recalling from the definition of importance that $P^*\,(t > \hat{t}) = 0$],

$$I_n^B(\hat{t}) = \frac{\partial g_s}{\partial q_n} = \left(\frac{\partial q_n}{\partial\lambda_n}\right)^{-1} \int_0^{\hat{t}} \hat{P}^* \frac{\partial M_s}{\partial\lambda_n} P\, dt \tag{35}$$

where now the importance function $P^*\,(t)$ obeys Equation 24 with source vector $h_a^+ = h_0^+\delta(t - \hat{t})$ and h_0^+ given by Equation 20. We can easily verify that, in this case, source vector h_a^+ in Equation 24, relevant to the importance function, accounts now for the final condition at $\hat{t}$, i.e., $P^*(\hat{t}) = h_0^+$.

From what we have seen above, the GPT methodology can given valuable information relevant to a given system, either from the point of view of performance (e.g., via sensitivity analysis for system availability optimization), or from the point of view of safety (by allowing an accurate evaluation of the system reliability).

The practicability of the GPT methodology is connected with the possibility of adopting the Markov chain model for system probabilistic analysis. We have seen in Chapter 3 how this model can be adopted even for large systems, due to the simplified structure into which the transition matrix can be reduced in practical cases and considering also the increasing memory capacity and calculational speed of modern computers. Further efforts and research may still allow further important simplifications applicable to real situations. A significant contribution in this direction is shown in Appendix B where a recent reduction method is illustrated, for the derivation of which heuristic concepts analogous to GPT theory have been used.

Another important aspect of the GPT methodology should be mentioned, concerning the case in which the hazard and repair rates of the system components can be assumed constant during the mission period $(0,t_F)$ and we are interested in evaluating perturbation changes, or sensitivity coefficients, of the type of Equations 33 or 34 respectively, i.e., relevant to time-dependent quantities. More precisely, in those cases in which reliability (or unavailability) perturbation changes or sensitivity coefficients are requested at different times $\hat{t}_l$, ($l = 1,2,\ldots$) during the mission period $(0,t_F)$.

With the above assumption, it results, in fact, that the importance function entering the GPT expressions needs to be calculated only once. This is so since in this case $P^*(t)$ depends, rather than on t, on the difference $(\hat{t}_l - t)$, where $t < \hat{t}_l$. Then, by the mere resetting of the final time t_l, the same importance function can be used for each different case.

IX. RELIABILITY UNCERTAINTY ANALYSIS

In reliability studies the data base (component hazard rates, repair rates, etc.) used for the estimates may be affected by considerable errors, sometimes within orders of magnitude. In these circumstances, it is essential to make an estimate of the propagation of these inaccuracies on the availability (reliability) calculations. For this problem there are a number of approaches.

Let us consider, for example, the system unavailability. Let us first expand it in terms of the system parameters α_i (representing hazard rates, repair rates, etc.). It results

$$\overline{A}_s(t) = \overline{A}_{s,0}(t) + \sum_i \frac{\partial \overline{A}_s}{\partial \alpha_i} (\alpha_i - \alpha_{0,i}) +$$

$$\frac{1}{2} \sum_{i,i'} \frac{\partial^2 \overline{A}_s}{\partial \alpha_i \, \partial \alpha_{i'}} (\alpha_i - \alpha_{0,i})(\alpha_{i'} - \alpha_{0,i'}) + \dots \qquad (36)$$

where $\alpha_{0,i}$ correspond to parameter (unbiased) estimates obtained from the data base available. To second order, we can write, assuming independent parameters α_i,

$$E\{\overline{A}_s(t)\} = \overline{A}_{0,s} + \frac{1}{2} \sum_i \frac{\partial^2 \overline{A}_s}{\partial \alpha_i^2} \, \text{var}(\alpha_i) \qquad (37)$$

The above expression illustrates how a correct (unbiased) estimate of the system unavailability generally differs from the value $\overline{A}_{0,s}$, obtained with the data base available, the discrepancy being proportional to the variances of parameters α_i and to the second order derivatives $\partial^2 \overline{A}_s / \partial \alpha_i^2$.

Likewise, we obtain the variance, var $(\overline{A}_s)$, which results

$$\text{var}(\overline{A}_s) = \sum_i \left(\frac{\partial \overline{A}_s}{\partial \alpha_i} \right)^2 \text{var}(\alpha_i) \qquad (38)$$

In this case, first order derivatives (sensitivities) are required to the estimate.

Once variances, var (α_i), of the system parameters are known from the data base, Equations 37 and 38 can be used for determining estimates and of its variance.

The derivatives (sensitivities) required in the above expressions can either be calculated by numerical direct methods, or they can be estimated making use of the GPT methodology, within the Markov chain representation, as seen in previous section.

APPENDIX A: GPT METHOD[7,8]

Let us consider a generic physical system described by a number of parameters p_j $(j = 1,2, \dots J)$ and by an N-component vector field $\mathbf{f}$ obeying equation (in vector notation)

$$\mathbf{m}(\mathbf{f}|\mathbf{p}) = 0 \qquad (A.1)$$

Vector $\mathbf{f}(\theta,t)$ describes the state of the system and generally depends on the phase-space coordinates θ and time t. Vector $\mathbf{p}$ represents the set of J parameters fully describing the system and entering into the calculation. Equation A.1 can be viewed as an equation comprehensive of linear, as well as nonlinear, operators and is assumed to be derivable with respect to parameters p_j and, in the Frechet sense,[9] to functions f_n.

Consider now a response of interest, or functional, Q given by the expression

$$Q = \int_{t_0}^{t_F} <\mathbf{h}^{+T},\mathbf{f}> \, dt \equiv \ll\mathbf{h}^{+T},\mathbf{f}\gg \qquad (A.2)$$

where $\mathbf{h}^{+T}$ is the transposed of an assigned vector function $\mathbf{h}^+$, t_0 and t_F represent given time limits, while brackets $< >$ represent integration over the space-phase, whereas the double ones $\ll\gg$ also over time.

In the following, we shall look for an expression giving perturbatively the change δQ in terms of the perturbations δp_j of the system parameters. In particular, sensitivity coefficient relevant to each parameter p_j will be obtained.

According to the GPT method, Equation A.1, or its linearized form (if dealing with a nonlinear problem), is heuristically interpreted as governing some density, more correctly, a (pseudo) particle field. The concept of importance [denoted as $\mathbf{f}^*$ (θ,t)] can then be introduced, corresponding to the contribution to a given functional due to the insertion of, say, a (pseudo) particle in a given phase-space position and at a given time.

Expanding Equation A.1 around a reference solution gives

$$\sum_{j=1}^{N} (H\mathbf{f}_{/j} + \mathbf{m}_{/j})\delta p_j + 0_2 = 0 \qquad (A.3)$$

where 0_2 is a second, or higher, order term, and where

$$\mathbf{f}_{/j} = \frac{d\mathbf{f}}{dp_j} \qquad (A.4)$$

$$\mathbf{m}_{/j} = \frac{d\mathbf{m}}{dp_j} \qquad (A.5)$$

Operator H is given by the expression

$$H = \begin{vmatrix} \dfrac{\bar{\partial}m_1}{\partial f_1} & \dfrac{\bar{\partial}m_1}{\partial f_2} & \cdots & \dfrac{\bar{\partial}m_1}{\partial f_N} \\[2ex] \dfrac{\bar{\partial}m_2}{\partial f_1} & \dfrac{\bar{\partial}m_2}{\partial f_2} & \cdots & \dfrac{\bar{\partial}m_2}{\partial f_N} \\ \cdots\cdots\cdots\cdots\cdots\cdots\cdots \\ \cdots\cdots\cdots\cdots\cdots\cdots\cdots \\ \dfrac{\bar{\partial}m_N}{\partial f_1} & \dfrac{\bar{\partial}m_N}{\partial f_2} & \cdots & \dfrac{\bar{\partial}m_N}{\partial f_N} \end{vmatrix} \equiv \dfrac{\bar{\partial}m}{\partial f} \qquad (A.6)$$

where by $\bar{\partial}/\partial f_n$ we have indicated a Frechet derivative.* Since in Equation A.3 the parameters p_j, and their changes δp_j, are assumed independent from each other, it must be

---

*   As a practical rule, these derivatives coincide with the ordinary derivatives if m contains only scalar operators. In case linear operators are contained, such as differential or integral ones, the usual rules of derivation can still be applied if these operators are formally treated as scalar quantities. So, if M is a given differential, or integral operator, then

$$\bar{\partial}(M\mathbf{f})/\partial f = M$$

For example,

$$\frac{\bar{\partial}}{\partial f}\left(\frac{\partial f}{\partial x}\right) = \frac{\partial}{\partial x}$$

$$\frac{\bar{\partial}}{\partial f}\left[\int dx \, K(x) \, f(x)\right] = \int dx \, K(x) \, (\cdot)$$

To note that a Frechet derivative corresponds to an operator.

$$\mathbf{Hf}_{/j} + \mathbf{m}_{/j} = 0 \tag{A.7}$$

which represents the (linear) equation governing the pseudo-density $\mathbf{f}_{/j}$. The source term $\mathbf{m}_{/j}$ is here intended to account also, via appropriate delta functions, of the initial conditions*
Consider now the response

$$Q_j = \langle\!\langle \mathbf{h}^+, \mathbf{f}_{/j} \rangle\!\rangle \tag{A.8}$$

Adopting the concept of importance to field $\mathbf{f}_{/j}$, if we weight with it space- and time-wise the source term $\mathbf{m}_{/j}$ (inclusive of delta functions accommodating initial conditions), this amounts to a result equivalent to the response Q_j, i.e.,

$$Q_j = \langle\!\langle \mathbf{f}^*, \mathbf{m}_{/j} \rangle\!\rangle \tag{A.9}$$

where $\mathbf{f}^*$ is the importance function-obeying equation[7]

$$\mathbf{H}^*\mathbf{f}^* + \mathbf{h}^+ = 0 \tag{A.10}$$

H* being obtained reversing operator H. This implies transposing matrix elements, changing sign of the odd derivatives, inverting the order of operators.
We can easily see that the sensitivity, s_j, of functional Q with respect to parameter p_j can be written

$$s_j = \langle\!\langle \frac{\partial \mathbf{h}}{\partial p_j}, \mathbf{f} \rangle\!\rangle + \langle\!\langle \mathbf{f}^*, \frac{\partial \mathbf{m}}{\partial p_j} \rangle\!\rangle \tag{A.11}$$

where the first term at the right-hand side represents the so-called, easy to calculate, direct term.
The overall change δQ due to the perturbation δp_j of system parameters can be written

$$\delta Q = \sum_{j=1}^{J} \delta p \left[\langle\!\langle \frac{\partial \mathbf{h}}{\partial p_j}, \mathbf{f} \rangle\!\rangle + \langle\!\langle \mathbf{f}^*, \frac{\partial \mathbf{m}}{\partial p_j} \rangle\!\rangle \right] \tag{A.12}$$

The above expressions, Equations A.11 and A.12 are quite general and can be applied to any response defined in a linear or nonlinear, time-dependent, or stationary, field. If the response is itself nonlinear, its linearization with respect to f should first be made and then the above methodology applied, or, which amounts to the same, such nonlinearity may be removed by augmenting the field by another variable. More explicitly, if $Q = <<L(f)>>$,

---

* To exemplify, consider equation

$$m(f) = -\frac{df}{dt} + af = 0$$

with initial condition $f(t_0) = f_0$. If we substitute it with the equivalent equation

$$m(f) = -\frac{df}{dt} + af + f_0 \delta(t - t_0) = 0$$

then we could have

$$m_{/j} = -\frac{\partial a}{\partial P_j} f + \frac{\partial f_0}{\partial P_j} \delta(t - t_0)$$

L(f) representing a generic, derivable, nonlinear scalar expression function of f, we consider the extra equation $y = L(f)$, where y represents the new variable, so that the new field is $\hat{f} = |\dot{\hat{y}}|$. The response results automatically linearized since it can be written

$$Q = \ll h^+, \hat{f} \gg \qquad (A.13)$$

with

$$h^+ = |1 \; 0 \; 0 \ldots 0| \qquad (A.14)$$

Second or higher order perturbation expressions can as well be derived following a similar procedure.

APPENDIX B: MARKOV MODEL REDUCTION

We have already mentioned that, in relation to large systems, Markov chain modeling can be complicated by the large number of states to be considered, so to exceed the general computer capacity. In order to extend the Markov approach to these systems, a simplification of the model can be made in those cases in which the instantaneous repair rate of components by far exceeds the hazard rate (which occurs in the majority of practical cases). The method we are going to illustrate has been proposed by Tagaraki et al.,[10] making use of a sensitivity methodology developed by Dieudonnè.[11] Here we shall illustrate this same methodology making use of GPT theory.

The basic idea is that of identifying, within the above (usually satisfied) hypothesis concerning the repair and hazard rates, those transitions between states due to component failures which are irrelevent to the unavailability value (assumed to represents the functional quantity to be evaluated). Suppressing these transitions by setting zero, the corresponding coefficients in the transition matrix, eventually results into splitting the set of all the system states into two subsets. That containing the starting state (usually, with index 1) represents the reduced Markov chain searched.

Identification of the possible candidates for these transition suppressions can be eased by use of a Markov transition graph. Consider, for illustration, the simple example in Figure 1, representing a one-out-of-three system. The system success state is 1 and the system failure states are from 2 through 8. The failure and repair rates of the ith component (i = 1,2,3) are indicated as λi and μi, respectively. It is also assumed that three repairmen are available. It may be of interest to explore the possibility of suppressing the failure transitions between states 2 and 5, 2 and 6, 3 and 5, 4 and 6, 7 and 8, or, more appropriately, to cut the graph (λ) edges (2,5), (2,6), (3,5), (4,6), (7,8), respectively, which would split the graph into two subgraphs including states 1,2,3,4,7 and states 5,6,8, respectively, as shown in Figure 1. The first subgraph would correspond to the reduced Markov chain searched.

In order that this reduction search is made possible we need a method which allows to evaluate the effect of these graph edge cuts with a limited effort. Making use of arguments based on GPT theory, we shall show here how such a method can be simply obtained.

Let us consider first the probability, $P_r(t)$, that at time t $[0 \leq t \leq t_F$ (mission period)] the system reaches a given state r. If we define the functional

$$P_r(\hat{t}) \equiv \int_0^{t_F} h^{+T}(t) \; P(t) \; dt \qquad (B.1)$$

with vector h^+ having components $h_n^+ = \delta_{nr}\delta(t-t)$ (δ_{nr} being the Kronecker symbol), we can write the GPT sensitivity expression of $P_r(t)$ with respect to a generic element $a_{kk'}$ of the transition matrix M_a (see Appendix A). It results

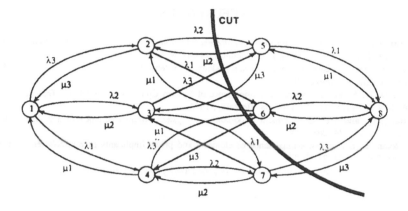

State	Components (UP or Under Repair)		
	1	2	3
1	UP	UP	UP
2	UP	UP	UR
3	UP	UR	UP
4	UR	UP	UP
5	UP	UR	UR
6	UR	UP	UR
7	UR	UR	UP
8	UR	UR	UR

FIGURE 1.   Example of Markov transition graph of a system with three components.

$$\frac{\partial P_r}{\partial a_{kk'}} = \int_0^i P^*_{(r)} \frac{\partial M_a}{\partial a_{kk'}} P \; dt = \int_0^i P_{k'} \, (P^*_{(r)k} - P^*_{(r)k'})dt \qquad (B.2)$$

where suffix (r) reminds that vector function **P*** depends on the state r considered at the final time t at which all its components vanish excepting P_r^* which assumes a unit value. Now, recalling the meaning of importance, $P^*_{(r)k}$ and $P^*_{(r)k'}$ represent the contribution to the final state r from state k and k', respectively. This means that, with the assumption made on the component failure and repair rates, if state r can be reached by repair, their value and, then, that of the above derivative, Equation B.2, will be much greater than if r can be reached by failure. Most of the probability changes stemming from the change of $a_{kk'}$ are then going to be effective only within the subgraph formed by the two states k and k' and by those reachable from these by repair. This in turn suggests that if we set $\delta a_{kk'} = -a_{kk'}$, i.e., we suppress the transition between the two states, or, equivalently, we cut the corresponding graph edge, this effect can be practically studied within such subgraph. If by proper analysis of the problem, it results that the effects on the transitions internal to it due to the edge cut are negligible, then we can extrapolate this conclusion, with corresponding negligible unaccuracy, to the original graph and proceed investigating on the next edge cut. This procedure continues hopefully until the graph is split into two noncommunicating subgraphs. That containing the initial state represents the reduced Markov chain model to be analyzed.

The above reducing procedure can be easily generalized if, rather than between failure and repair states, we would more generally distinguish between smaller value and larger value transition ones.

REFERENCES

1. Birnbaum, L. W., *On the Importance of Different Components in a Multi-Component System. Multivariate Analysis II*, Krishaiah, P. P., Ed., Academic Press, New York, 1969.
2. Lambert, H. E., Fault Trees for Decision Making in Systems Analysis, Report UCRL-51829, University of California, Livermore, 1975.
3. Barlow, R. E. and Proschan, F., Importance of System Components and Fault Tree Analysis. Report ORC-74-3, University of California, Berkeley, 1974.
4. Zhang, Q. and Mei, Q., Element importance and system failure frequency of a 2-state system, *IEEE Trans. Reliab.*, 24, 208, 1985.
5. Jackson, P. S., On the s-importance of elements and prime implicants of non-coherent systems, *IEEE Trans. Reliab.*, 32, 21, 1983.
6. Gandini, A., System reliability analysis via generalized perturbation theory methods, *Trans. Am. Nucl. Soc.*, 53, 265, 1986.
7. Gandini, A., A generalized perturbation method for bi-linear functionals of the real and adjoint neutron fluxes, *J. Nucl. Energy*, 21, 755, 1967.
8. Gandini, A., Generalized perturbation theory (GPT) methods. A heuristic approach, in *Advances in Nuclear Science and Technology*, Vol. 19, Lewins, J. and Becker, M., Eds., Plenum Press, New York, 1987, 205.
9. Linsternik, L. A. and Sobelev, V. J., *Elements of Functional Analysis*, Frederick Ungar, New York, 1972.
10. Takaragi, K., Sasaki, R., and Shingai, S., A method of rapid Markov reliability calculation, *IEEE Trans. Reliab.* 34, 262, 1985.
11. Dieudonnè, J., *Foundation of Modern Analysis*, Vol. 10, Academic Press, New York, 1969, 295.

Section III
Internal Damage Risk Causes

INTRODUCTION

Preliminary to whatever action regarding an industrial plant (especially a high-risk one), including design of a part, construction of a component, testing, operation, licensing, etc., should be a sufficient knowledge of the process involved and of the main safety aspects of the whole plant.

While in the design of plants or equipment not handling high-risk matters it is justifiable to consider a peculiar aspect without a knowledge of the whole, in the opposite case that knowledge is fundamental, because the safety-oriented mind or the safety-feeling must act as a continuous filter in the approach to whatever problem of design, operation, or control.

This might be considered as the most relevant general safety criterion, an essential presupposition for the "true" safety of the plant. When this criterion is accomplished, the most relevant aspect affecting the safety of the plant is the capability of the designer to conceive the plant itself in safety, well understanding the processes involved and their possible modes of evolution and identifying all causes, internal to the processes or external to them, that could be cause of damages, thus leading to harmful releases to the environment, so as to adopt the most suitable means of defense of the integrity of the plant.

Due to the scope of this book, only design criteria of a general kind have been examined, leaving design criteria specific of peculiar processes apart.

Some emphasis has been devoted to the analysis of the causes of damage for the plant, on which the integrity of the plant itself (and therefore the nonrelease of harmful substances and the nonrisk to man) depends. What has not been considered in this book are the causes of damage (and of releases and of risk) due to mistakes in the design or in the operation: it is clear that if the designer has failed in computing the thickness of a vessel or a piping even normal operation conditions could be responsible of an accident. What is here assumed is that the design of the plant handling high-risk substances has been performed by qualified people, using qualified computing tools, under qualified procedures (application of quality assurance procedures), suitably checked, and the same standard of quality (or qualification) has been adopted in selecting materials, component construction, assembling, erection, and testing of the plant.

In other words, the high quality of the project is assumed as certain, and to such aspect the licensing and control authorities must devote the maximum care. What is analyzed in Sections III and IV are the various types of "internal" and "external" "special" events, not regarding normal operation conditions for the plant, whose applicability shall be considered by the designer, the operator, and the licensing authority. A brief introduction to such special events is in the following.

In general, those industrial activities (except for those of military character) which present a risk to man can be divided into two wide categories: (1) production activities — storage and (2) transportation activities — usage.

The first category develops according to the following specific activities:

1. Production-storage of substances and manufactured goods (e.g., chemical plants)
2. Production-storage of information (e.g., research laboratories)
3. Production-storage of energy (e.g., thermoelectrical, hydraulic, and nuclear plants)

In the second category there are the same specific activities which correspond to the production-storage, with the addition of the activity of transportation of people, which is perhaps the activity which constitutes the greatest contribution to the overall risk.

As already specified, in this book reference is made exclusively to industrial activities related to the production-storage of inflammable, explosive, toxic, or otherwise dangerous substances which may be present in industrial production plants.

When considering an activity of production-storage of inflammable, explosive, toxic, or otherwise dangerous substances, it is possible to identify two categories of events which contribute to the risk value of this activity and which exhaust the whole series of problems: (1) internal events — those events which originate from an internal malfunctioning in the plant and (2) external events — those events which originate outside the plant (as external forces) and which can cause malfunctioning at the plant.

The results of these two types of events cause malfunctioning in the plant systems which, if not reducible to limited consequences, could cause, in turn, damage outside of the plant itself.

In the first cause (internal events) therefore, there is a direct risk which is born within the plant; in the second one (external events) it is possible to speak of risk induced by other causes against which an industrial plant must be protected. In particular, external events which may be harmful for the industrial plant and become responsible of its damage and therefore of higher risks are essentially of two types:

1.   Natural events, including earthquakes, whirlwinds, and floods
2.   Events caused by human activity, including aircraft crashes, explosions, fire from the outside, presence of toxic, or otherwise dangerous gases originated out of the plant

As a general criterion, every industrial activity, so as not to cause relevant risks to the population and the environment, should either be located at sufficient distance from all other activities, or should be adequately protected against external events.

The necessity of assessing the risk from internal events remains evident and the criteria which will be described in the following chapters represent a basis for such an assessment.

During the operation of an industrial plant, internal events which could be cause of risk to the population and the environment are

1.   Fire of substances inside containers and equipment
2.   Explosion of substances inside containers and equipment
3.   Damage to containment systems with release of substances into the external environment, with eventual damage by contamination or secondary production of inflammable compounds, explosives, etc.

Such events can be present either separately or in combination; in general, the initial cause is the loss of balance in the physical-chemical properties of a process or in its energy balances or loads. In practice, such a loss of balance could be caused by a breakdown or degradation in a system, human error, or unforseen conditions.

To avoid the occurrence of the above-mentioned conditions of loss of balance (prevention) and/or to limit the gravity of the consequences (mitigation) they could produce, suitable engineered plant systems may be necessary.

In particular, there are two classes of potentially dangerous conditions for industrial plants:

1.   Dependent on the chemical substances present in the process
2.   Dependent on breakdown of equipment

The conditions of danger due to chemical properties events which depend on the physical-chemical properties of the materials present in the production process or — in general — in the plant. Such conditions must be taken into account both for the single substances as well as for those in combination. The following are part of this first class of potentially dangerous conditions:

1. Reactions whose final effect is represented by one of the three events of risk to the population and the environment or by a combination of the three (fire, explosion, and release of substances)
2. Toxic reactions, i.e., the formation of byproducts which are noxious to the population

By carrying out systematic research into the properties of a single substance or a combination of chemical substances present in the plant, it is possible to show all the combinations of danger dependent on chemical types. The conditions of danger caused by process equipment are generally events due to loads which exceed that foreseen in the design of such equipment. In particular, it is a question of overloading caused by fluid pressure or structural overloading. The reference magnitudes, in this case, are pressure, temperature, and velocities of moving parts of the equipment.

In Sections III and IV the main aspects to be considered for the evaluation of the dangerousness of the internal and external events for the specific plant, are considered.

It is important to recall that while local design rules and regulations often represent an exhaustive reference for the design of components of a plant during normal operation (including start-up and operational transients), for what concerns the behavior of the plant under accidental conditions and even for what concerns the selection itself of the internal and external events to be considered to guarantee the capability of withstanding of the plant, there is very often a lack of rules and regulations, which increases the responsibility of the designer.

Chapter 6

GENERAL SAFETY-RELATED DESIGN CRITERIA

Sebastiano Serra

TABLE OF CONTENTS

I. INTRODUCTION

The safety criteria for an industrial plant should follow two basic guidelines: (1) prevention of accidents and (2) mitigation of consequences whenever an accident takes place in spite of the preventive measures adopted.

Preventive measures are a fundamental issue in design as they render almost negligible the probability to cause undesired conditions within industrial plants.

II. BARRIERS AND PREVENTION

Barriers preventing dangerous situations within industrial plants shall be identified on the basis of the accident configuration pattern. In practice, on the basis of the process characteristics, the components and phenomenological conditions, as well as their position within the process that lead to the loss of the barrier, are identified. Thus, the barrier shall be intended in its widest sense; it is always represented by a precise physical component, but if the initiating event causes the loss of the physical barrier, the latter shall be identified upstream of the physical component, e.g., in a control or measuring system, that will therefore act as a barrier for the initiating event. For example, in a process system including a harmful substance, system components (vessel, pumps, valves) are the barrier to the diffusion of that substance. However, if for a given primary event, one of the components loses its containment function because of a failure of a control system, then the barrier function is no more linked to the sole physical component but to the system that had to ensure the functionality of the component. When the circuit pressure opens a safety valve because of a primary event, the barrier is lost through that valve and its malfunctioning might determine a continued emission of the harmful substance to the environment, if the measuring system does not allow closing of the open circuit by the operator. In this case, the measurement circuit and all components allowing the interception, operator included, became basic elements of the barrier. Thus, we can say that an accident occurs when due to the initiating event said element causes the loss of all barriers available to confine the fire, explosion, or release of substances to the external environment.

General design criteria are precise rules which the designer shall adopt to maintain barrier characteristics and thus withstand both the initiating and the consequential events.

Preventive measures should be implemented through several ways. The following are the most important:

1. High quality of the plant design and construction; adoption of design criteria for the systems, structures, components, and materials which would impair safety and protection of workers, population and environment in case of their failure.
2. Implementation of a "quality control" program to ensure that all the design and construction rules and procedures are enforced and documented.
3. The design should carefully take into account the safety-related systems, internal stresses (linked to the process), and stresses due to events occurring outside the plant such as earthquakes, floods, tornados, and extreme environmental conditions.
4. Systematic inspection and test of the safety-related components at the shop after installation and periodically during operation, to ascertain their good performance.
5. Adoption of control and protection systems maintaining process parameters within the operating range and automatically stopping the plant whenever any working troubles should occur.
6. Control of the main barriers during operation; sampling and monitoring systems controlling the integrity of the circuit containing the process fluid in question should be installed to this purpose.

7. Preparation of local panel boards to be used whenever the main control room is not available, in order to use the safety-related systems.
8. Training of the operating staff on similar plants or simulators, in order to have qualified operators for both normal and emergency situations.
9. Spreading of the operating experience acquired in all facilities to ensure that in case of troubles or anomalies capable of provoking them it is possible to timely act and avoid that the same problem occurs in other facilities.
10. Independent control and operation by controllers especially appointed to this purpose.

The above-mentioned design characteristics and preventive measures could be the principal safety guarantee of industrial plants. However, apart from preventive and protective measures, it should be possible to hypothesize the occurrence of accidents and install safety systems capable of supplying a second level of prevention.

This second level is achieved by trying to confine the intiating event and any events deriving therefrom throughout the whole plant. This entails limitation of the initiating event and availability of safety systems based on well-defined criteria.

In particular, the safety systems should have the following characteristics:

- High intrinsic reliability of the whole system, of single systems, and their components.
- Redundancy of subsystems or components, to be sure that the systems perform their functions also in case of failure of unavailability of one subsystem or component.
- Physical separation of subsystems or redundant components and diversification of signals that actuate the safety systems, to avoid common mode failures and increase reliability of the system themselves.
- Power supply always available both for energy and control needs.
- Water storage always available for the hydraulic safety systems.
- Functional testing of components and maintenance services if necessary at short intervals.

A. Redundancy and Common Mode Failure

One of the requirements to have highly reliable systems, is redundancy. This method uses multiple subsystems that carry on a given safety function even in the case of loss of a single subsystem; in this way, the initiating event is but a simple plant anomaly. The redundancy has however a limited capacity to increase the system availability due to common mode of failure.

We will shortly describe the common mode failure and the way to deal with it.

The common mode failure can be defined as a series of failures related to one another that cannot be considered as separate events. Multiple failures are common or dependent modes, since they derive from a single initiating cause. Possible individual initiating causes are common characteristics, common process, common environment, and common external event.

Epler was the first to stress this aspect of failures in 1968. His investigation was based on a small sample of data, but offered interesting conclusions: having two redundant units in a system, with an accidental failure probability which is lower that of the common mode, then there is no need to use three of those units except for reasons of coincidence or convenience.

More accurate analysis identified the mechanisms that cause the potential common failure:

- Design deficiencies
- Nonobservance of manufacturing, assembling, and quality control specifications
- Errors during tests, maintenance, and repairs

- Human errors
- Unexpected environmental conditions
- Failures or degradations due to another failure; the failure is produced outside the system

For instance, if we have two components A and B in parallel, the probability of failure of the component A is P(A), and the probability of failure of the redundant component B is P(B). P(AB) is the failure probability of both A and B components; it always follows that:

$$P(AB) = P(A) \cdot P(B/A) \tag{1}$$

where P(B/A) is the failure probability of B related to the failure of A.

When A and B are totally independent, P(B/A) = P(B) and if for example:

$$P(A) = 10^{-3} \quad \text{and} \quad P(B) = 10^{-3}$$

then

$$P(AB) = P(A) \cdot P(B) = 10^{-6} \tag{2}$$

that is, the product of the failure probabilities of each component. However, if A and B use the same operation-base physical law, are installed by the same person, are manufactured by the same company, are in the same environmental conditions, are maintained by the same personnel, etc., then P(B/A) probability is different from P(B).

With the above-mentioned limit conditions P(B/A) can be very high (≈ 1 unit), therefore, if we apply Equation 1 we have P(AB) = $10^{-3} \cdot 1 = 10^{-3}$.

Thus, Equation 2 is no more applicable if the events are not independent and may fail for a common cause.

$$P(AB) \neq P(A) \cdot P(B) \tag{3}$$

Hence, in order to have highly reliable safety systems it is necessary to eliminate the potential common mode failure to the maximum possible extent.

If this type of failure plays a major role in the development of accidents, it is advisable to qualitatively analyze a strategy to use in industrial plants rather than describing a method to quantitatively assess the influence of the common mode failure.

From the above list of mechanisms causing the potential common failure, it is evident that unexpected environmental conditions, failures or degradations due to another failure, or failures outside the system can be generally identified and eliminated by means of an accurate reliability analysis.

In particular, unexpected environmental conditions (temperature, excess of humidity, dust, electromagnetic interferences, vibrations, floods, fires, etc.) can be singled out through an analysis of the location of the system components; the most efficient means of protection for these types of common mode mechanisms is to locate the component in separate rooms and far from another.

Failures or degradations due to another failure or to a failure outside the system can be identified by means of the functional reliability analysis, taking into account auxiliary safety systems (controls, power, environmental conditions) which act as a support to the correct system operation.

Diversification is the last efficient means, to be used with caution after making a correct reliability analysis, to eliminate common mode failures. This means should be used with great caution as it could cause serious complications to the safety systems, because elimination of the common mode failures could imply a net loss of reliability due to the increased number of casual failures. Thus, diversification should be applied to particular circuit elements (initiating signals or special components such as valves or pumps) only.

All the above-listed methods do not allow complete elimination of the multiple human errors which could cause the common mode failure, such as:

1. Design errors.
2. Errors made during testing, maintenance, and repair of failed components.
3. Errors made in the operation of the plant (e.g., improper calibration, nonobservance of operating procedures, procedural errors).

Some useful strategies to reduce the common mode failure for this kind of errors are available.

As far as the designing is concerned, the most important aspects to take into account are the following:

1. Administrative controls: procedure and supervision of the design, specifications, manufacturing, and installation considering that errors during tests and maintenance might jeopardize good plant operation; reliability analysis as integral part of the project strategy.
2. Fail-safe design: the design should provide for safe plant operation with a multiple failure avoiding them rather than design for a single failure.
3. Quite flexible intervention modalities, adequate to testing and maintenance conditions; it should be possible to install separate sensors for each barrier.
4. Safe, standardization, and simplicity; design modifications, complexity, and difficult procedures should be avoided; the equipment characteristics in normal operation conditions should be far below the maximum values of the parameters inducing high stresses in the equipment.
5. Design review: independent and thorough review of the design, including a reliability analysis.

As to the operation of plants, the principal factors that allow reduction of the common mode failure are as follows:

1. Administrative controls of the plant management to ensure that the design reliability is not endangered during the operation and life of the plant, i.e., plant failures should not affect the safety-systems.
2. Maintenance procedures to prevent failures or repair them. Every maintenance activity aims at improving the system reliability but entails a high potential of common mode failure due to the human factor. For this reason, some expedients are required: clear and easy maintenance procedures, tests after maintenance to be carried out with special care for stand-by systems, and routine maintenance of the subsystems at different times.
3. Testing procedures aiming at identifying common mode failures as well as single failures, and at avoiding the introduction of common mode failures due to human errors during tests; simple, unambiguous, explicit, and understandable procedures, subsystems testing made at different times to avoid that the barrier protective measures are bypassed by the test itself, and immediate repair after failure of the equipment under test.

4.   Operating procedures capable of maintaining the design safety margins for the barriers.
5.   Monitoring of reliability features throughout the life of the plant, recording of the failure data relating to all the activities (operation, maintenance, tests) to reduce the possible occurrence of failures.

From the foregoing considerations on industrial plants, it follows that safety probabilistic studies including reliability analysis are a good way to reduce the common mode failure. The reliability analysis appears to be one of the best ways to know the industrial plant process; it provides a more systematic framework which cannot be obtained with the sole use of individual experiences or a deterministic approach.

III. BARRIERS AND MITIGATION

Mitigation inside the plant is provided by a series of barriers that avoid an internal event to be seriously dangerous for the environment and the population in the vicinity of the same plant.

In fact, after passing all preventive barriers, the internal event would have direct effects outside by spreading the fire or explosions or through the release of toxic substances at a concentration which would be harmful for the environment and the population health.

Thus, mitigation aims at avoiding all this by means of devices capable of limiting the extent of such phenomena.

This function may be achieved through containment systems.

As to fires, the containment role is played by firebreaks, the distance from the population, and the environment as well as the implementation of emergency measures.

As for explosions, the most efficient mitigating means is represented by the distance effect or by suitable barriers.

Regarding release of toxic fluid substances, physical containment barriers and suitable emergency plans should be provided.

A. Containment

Containment is the most important barrier to the release of harmful substances to the external environment.

The study of efficient containments is based on the accurate analysis of possible release pathways, after passing all preventive barriers.

As a general rule, in a process system the overpressure safety systems are likely to lose their ability to confine the dangerous fluid. Thus, consideration should be given to relief valves, isolating valves, depressurization systems, drainage and filtering systems, and to every connection of the process with the external environment in general. A containment system can be provided either by a single container enclosing the whole process or by a series of containments apt to mitigate the release to the environment so that no damage occurs outside the plant.

The design of such containments should also meet certain safety requisites; some of them have been already described, but they are worth a more detailed description because of the precise function of containment.

Containment is basically composed of a physical structure and of the auxiliaries; the latter are of major importance for its correct function and contribute to limit possible releases of dangerous substances to the environment. In particular, auxiliaries play a major role for the safety of the systems described in Section II.A and should be designed on the basis of the previously listed criteria. However, the main difference both for containment and its auxiliaries is linked to the working conditions of the whole containment-auxiliary set. They should be checked against accidental conditions and designed in such a way as not to lose their functionality vis-a-vis the source of the primary event and the following loss of barriers.

The design of the containment system, should therefore establish the containment leak tightness or its retention capability to the release of harmful substances under expected accidental conditions. This characteristic is a function of the theoretical release amount and of those limit values which are acceptable outside the plant in case of accident. Limitation of prohibitive conditions (excessive pressure or temperature) in the physical structure of the containment and the conservation of both leak tightness and retention level shall be entrusted to the containment auxiliary systems.

It is extremely important to identify every possible bypassway of the containment system that existed even before the accident status, and the plant capability to minimize the probability and consequences of such events.

In practice, also a simple bubbling tank of a potentially dangerous process fluid shall be designed according to the above criteria. No factor shall be neglected in the identification of possible ways of loss of the mitigative barrier, in order to maintain a high degree of reliability by means of simple devices.

This approach is based on simpler deterministic factors, through the analysis of limiting accidents. In this case the reliability analysis (probabilistic approach) is linked to accidental events and to the behavior of auxiliary systems and structures, and is therefore more complex than the classical fault tree analysis. Nonetheless, the systematic probabilistic approach is efficient as well. However, the event tree technique is extremely useful to deal with this aspect of mitigation.

The emergency intervention actions for mitigation of the accident consequences should be studied and set into procedures to give the operators a clear indication of what preventive actions should be taken in these cases, as well as provisions to enforce within short periods. Emergency mitigating provisions constitute the last barrier along with the external emergency plan, in order to reduce the accident impact on the external environment; they should be carefully studied during the design phase and utilized for the operation of the industrial plant.

Chapter 7

THERMODYNAMIC DAMAGE CAUSES — SAFETY ANALYSIS

G. Lelli

TABLE OF CONTENTS

I. TYPICAL PHENOMENOLOGIES IN CONFINED SYSTEMS: REVIEW OF SOME THERMODYNAMIC CONCEPTS

Let us consider, for simplicity, a closed thermodynamic fluid system exchanging heat only with the environment, in steady-state conditions; in such a situation, we have a perfect balance between the input and the output thermal power. Should, for any cause, from a certain time onward, the input thermal power become greater than the output thermal power, a transient would occur, with an increase of the thermodynamic system internal energy; this would inevitably cause a rise of temperature inside the confined fluid and, consequently, an increase of the pressure exerted on the containment structures. These structures would tend to warp progressively, in such a way as to increase the internal volume available for the fluid. If we consider the whole thermodynamic system (fluid + containment structures) and do not take into account the negligible external work connected with its variation of volume, the first principle of thermodynamics is written:

$$dU_{tot}/dt + Q_e = Q_{f,in} - Q_{f,out} \tag{1}$$

where the terms on the left side are, respectively, the derivative, with respect to time, of the total internal energy and the thermal power given to the environment by the system (the environment is supposed to be cooler than the system), while the terms on the right side are the total thermal power given to the fluid and the thermal power removed from the fluid by heat exchangers included in the system. From Equation 1 we can see that, if a positive thermal unbalance occurs, the confined system reacts, partially (but usually prevailing) by increasing the internal energy of the containment structures and the specific internal energy of the fluid (the system mass remains the same) and, less significantly, by increasing the heat losses to the environment. The rate of change of the pressure inside a closed system, where the specific energy and the temperature of the fluid increases, depends on the fluid state as well as on the capability of the containment structures to be deformed. Since liquids are almost incompressible, a slight increase in temperature, at constant volume, is sufficient to cause a considerable increase in the pressure; however, a slight deformation of the containment structures (causing an increase of the internal volume of some per thousand) appreciably reduces the rise in internal pressure. Instead, the behavior of gases and super-heated vapor approaches, as temperature increases, the perfect gas behavior and, consequently, a considerable increase in temperature is required to have a valuable rise in pressure; the lower the fluid specific volume is, the greater this increase in pressure results. Finally, the behavior of two-phase mixtures is intermediate and depends mainly on the nature of the fluid and the value of the specific volume of each phase.

Considerable overpressures inside confined systems can also be connected with flow transients of the moving fluid; in such cases, in fact, variations of pressure occur, which tend to propagate quickly in the whole fluid system in form of compression and rarefaction waves of the fluid. This is the well-known phenomena called "water hammer", in its widest meaning.

On the grounds of the operating experience, it was found that water hammer can be caused by the following actions: fast operation of valves, sharp variations of pumps speed, entrainment of liquid plugs into steam flows, starting of pumps in partially empty lines, delayed opening of stuck on-off valves, separation of liquid columns inside pipelines, quick condensation of bubbles, and steam pockets inside loops. These pressure waves have peak values up to 10 MPa and can cause malfunctions and failures (of valves, measurement instrumentation, control, and protection systems) which are not always quickly detected and easily located. Moreoever, a water hammer causes significant dynamic stresses in the components of a pressurized loop and in the constraint and support systems as well, because of the deformation of the piping caused by the pressure wave.

II. CONSEQUENCES OF AN INCREASE IN PRESSURE IN CONFINED SYSTEMS

An increase in internal pressure causes a progressive deformation of the containment structures; however, as well known, this deformation cannot continue, in any real material, beyond a typical limit, since a loss of strength occurs and the containment system breaks down ejecting the confined fluid and, in some cases, parts of the pressure boundary. Failure of the containment structures, in consequence of pressurization, can be caused by plastic instability, elastic-plastic instability, and brittle fracture. This last is the most dreadful, because it propagates inside welded structures with a speed up to a 1000 m/s, generally without being preceded by external warning signals.* The direct consequences of a failure of the containment structures are connected with the nature itself of the ejected fluid. Generally, the pressure and/or the temperature of the fluid will be high; in high-risk plants the fluid system may include toxic, corrosive, and flammable substances (as in chemical industry), or radioactive substances (as in nuclear industry).

Furthermore, a failure can indirectly damage other plant components and systems, because of the possible effects, connected with the break, such as pressurization of rooms, jet forces, floods, and blows from shore pipes.

The economic consequences of a failure are, generally, mainly connected with the loss of plant production, given that the plant will have to be operated at reduced power or to be shut down for some time. Also, the costs for repairs of components and systems and decontamination of the work areas inside the plant and, in the most serious cases, of the surrounding areas, do not have to be undervalued. Finally, recent serious accidents, occurred in high-risk industrial plants in various countries, clearly showed the existence of other kinds of damages, besides those listed before, which burden the whole community and are hardly valuable in economic terms (death or permanent infirmity of people, temporary or definitive loss of houses, territory use, and products and manufactured articles no more marketable).

It follows that it is necessary to plan a series of suitable devices, able to prevent risky events as much as possible (e.g., overpressure inside confined systems), and moreover warrant that damages would be limited within acceptable values in the case of an accident. Such devices are the required qualification for the safeguard and the integrity of people, plant, and environment.

In the following paragraphs, the most important design solutions adopted to pursue these two aims are examined.

III. SIMULATION OF COMPLEX THERMOHYDRAULIC SYSTEMS

Unbalances between input and output thermal power are generally due to failures or malfunctions of instrumentation or mechanical components (regarded as the initiator events) leading to dangerous situations inside the confined fluid system; in such situations, excessive overpressure inside the system or loss of integrity of the containment structures could occur with harmful consequences for the environment, personnel, and population if the contained substances are toxic, radiotoxic, or inflammable.

During design and safety analyses, all the possible failure initiator events and the accident evolutions have to be considered in order to verify that the strength limits of the components are not exceeded and poisonous confined substances are not released beyond the established limits.

The analysis of the evolution of an accident has to be carried out in the time domain. That is particularly true in the case of complex systems, where numerous interfaces exist

---

* Leak before break.

among components and subsystems, and when the intervention of active emergency systems is expected. In this last case, delays of actuation of the emergency systems can involve significant influence on the accident dynamics.

For complex systems, the analysis of the thermohydraulic behavior of the system components and of the whole system, during a transient, becomes an essential tool for the definition of the mechanical design features and functional characteristics, for the analysis of the components performances under severe operating conditions, for the identification of the accidental and transient events that are critical for the integrity and safety of the plant, and, finally, for the determination of the actual plant safety margins with respect to such events. For high-complexity plants, such an analysis requires the use of suitable computation systems, able to simulate the thermohydraulic behavior of the whole plant and of its main components as well. To this end, computer programs have been developed, which may be divided in two categories: component programs and system programs.

Component programs are mostly aimed at design purposes and local analysis of specific phenomenologies, which, in some cases, can be of interest to plant safety. Such programs are therefore able to provide a very detailed and deepened description of the thermohydraulic behavior of single system components, and allow to point out those phenomena which influence the component behavior significantly.

The aim of the system programs is the analysis of the whole plant behavior and the thermohydraulic and thermodynamic evolution of the several parts of the plant itself; also, they describe the accident dynamics and the safety systems effectiveness. In order to do this, the models adopted in the system programs to describe the single components behavior aim at simulating only those phenomenologies which are important to represent the component macroscopic behavior and the interactions among the different components of the system, as well as the phenomena that are of interest for safety.

Hereafter, only system programs will be considered, except some cases explicitly indicated; typical problems connected with the use of such programs in the safety analysis of complex thermohydraulic systems will be examined.

A. Typical Components

From a functional point of view, in a thermohydraulic system we can distinguish the following main components:

- Heat source (boiler, chemical reactor, nuclear reactor)
- Heat sink (steam generator, condenser, heat exchanger)
- Expansion elements (valves, turbine, pressurizer, flashing tower)
- Regulation/check elements (valves, floodgates, etc.)
- Fluid driving elements (pump, blower, compressor, etc.)
- Transfer and confinement systems (pipes and tanks)

Some functional parameters of the system components can influence the transient dynamics; therefore, in order to obtain a computer program output sufficiently close to the real system behavior, the system program models will have to provide the correct simulation of such parameters. Regarding heat source and sink components, the crucial parameter is the input/output system thermal power. This quantity is often a function of a heat transfer coefficient (systems with heat flux not assigned), that becomes the critical parameter for the evaluation. When heat flux is imposed, also if the thermal power given to the fluid is independent of the heat transfer coefficient, the evaluation of it plays a fundamental role in determining the thermal conditions of the heat transfer surfaces. The exact calculation of the heat transfer coefficient is generally essential for a correct prediction of the transient behavior of the main thermohydraulic quantities such as pressure, enthalpy, and temperature.

Since the heat transfer coefficient strictly depends on the fluid characteristics, the right calculation of such a coefficient implies a correct evaluation of the heat transfer mechanisms (subcooled boiling, bulk boiling, burnout, etc.) and, consequently, of the flow regime which is established inside the channels involved in the thermal exchange. In general, we can assert that the main error source in the calculation of the heat transfer coefficient arises from the uncertainties associated with the evaluation of the physical and thermodynamic conditions of the fluids.

Regarding the valve component, as the expansion element, the most important parameter, for a correct prediction of the plant transient behavior, is the flow rate through the valve as a function of pressure and fluid characteristics upstream the valve itself.

An overestimate of the steam flow rate through a safety or relief valve involves an underestimate of the pressure inside the thermohydraulic system which the valve belongs to.

A right evaluation of the flow rate through the valve is a direct consequence of the simulation of the real dynamic behavior of the valve (opening/closing times, opening/closing setpoints) as well as of the adoption of suitable flow models, able to take into account variations of pressure and fluid characteristics (single phase or two phase with variable quality). The differences between the calculated flow rates and the experimental or experienced flow rates are mainly due to the inadequacy to simulate correctly the relationship between the flow rate and the fluid characteristics.

In system computer programs, it is often possible to obviate this inadequacy by implementing valve models having a flow area as a function of the upstream two-phase mixture quality.

Some critical phenomenologies of the "pressurizer" expansion element simulation involve similar problems; particularly, phase separation models (liquid and vapor), liquid entrainment models (by vapor), and bubble rise models (inside liquid mass) can lead to wrong evaluations of the two-phase mixture quality in the upper region of the component. Since the relief valves are located just in that area, an error results, with respect to the real case, in the calculation of the discharged fluid quantity.

Another crucial aspect about pressurizer is the correct evaluation of the thermal exchange between the walls and the fluid and the condensation-evaporation phenomena at the liquid-vapor interface. In fact, such processes modify the internal pressure of the component and cause pressure variations into the whole fluid system, which the component is connected to.

Finally, regarding the pump component, the moment of inertia, the head-flow curves over the four quadrants, and the pump performance characteristics under degraded two-phase conditions are the essential functional parameters in order to have a correct prediction of plant behavior, during a transient or an accident. The moment of inertia of the pump is the parameter governing the time delay required by the pump to stop, after a loss of power supply.

The head-flow curve, extended to the four quadrants, has an essential significance for the prediction of the plant thermohydraulic behavior, during a severe accident with a quick depressurization. As a simple example, let us consider a pressurized loop with a high temperature fluid, where a large break occurs upstream the circulation pump. Immediately after the break, a condition is established, called "energy dissipation condition", with the pump running in the positive direction and the mass flow moving in the opposite direction, because of the position of the break.

In order that the system computer program can simulate such an event, a suitable model correlating the pump head with the pump speed, in reverse flow condition, must be provided.

Finally, when a pump, designed to work in single-phase conditions, operates in two-phase conditions, the head-flow curve, as a function of the flow quality at the pump suction, has to be considered, in order to take into account the pump performance degradation in such conditions. Consequently, a suitable model able to simulate such a modification of the characteristic curve must be implemented in the computer program.

B. Simulation Problems

The simulation, by computer program, of a complex thermohydraulic system requires first of all the knowledge of a large amount of data about the plant and the single components. These data are all the geometrical dimensions useful to define flow areas, volumes, relative components arrangement (distances, elevations, inclinations), hydraulic characteristics (roughness, hydraulic diameters), thermal characteristics (heat transfer surfaces, thickness, heat sources distribution), as well as all the informations about materials (thermal conductivity, density) and the system boundary conditions (external temperature and pressure).

The analysis of the thermohydraulic behavior of a system consists in the evaluation, instant by instant, of the flow conditions among the various components and the thermodynamic conditions inside them. Obviously, such an analysis is based upon the solution of the mass and energy conservation equations applied to all the control volumes representing the system (often coincident with the physical components) and the momentum conservation equation applied to the lines connecting the control volumes.

Regarding the essential aspects of the conservation equations and the prominent heat transfer mechanisms (see Reference 1).

Here we want to recall only some particular components that require, for a suitable simulation, specific data, not always known or available because they are often unessential in a "traditional" design. Particularly, some crucial aspects of the simulation of pumps and valves, in transient thermohydraulic analysis, will be pointed out. Regarding the pump component, the importance of having the characteristic curves over the four quadrants has already been emphasized. Such curves, provided by the constructor, are obtained experimentally, and define pump head and torque as functions of volumetric flow-rate and pump speed.

A typical set of curves over the four quadrants is shown in Figure 1. In order to make their use easier, the same curves can be rewritten as homologous curves giving the head fraction* and the torque fraction as functions of pump speed fraction and volumetric flow rate fraction. A typical set of homologous curves giving the head and the torque, is shown in Figures 2 and 3 respectively.

In order to simulate cavitation and degradation of pump performances in two-phase conditions, two-phase homologous curves, giving head and torque, have been developed.

The experimental data show that total degradation conditions occur when the void fraction of the mixture is within the range 0.2 to 0.9, while, out of this interval, the pump provides some head, variable between the nondegraded conditions value (single-phase fluid) and zero (two-phase fluid). The calculation of pump head and torque is therefore accomplished by means of the following equations:

$$H = H_{1\phi} - M_h(\alpha) \cdot (H_{1\phi} - H_{2\phi}) \tag{2}$$

$$T = T_{1\phi} - M_t(\alpha) \cdot (T_{1\phi} - T_{2\phi}) \tag{3}$$

where H = head, T = torque, 1ϕ = single-phase value, 2ϕ = two-phase, fully degraded value, $M(\alpha)$ = multiplier, and α = void fraction.

Typically, the multiplying factor M varies between 0 and about 1, for void fractions from 0 up to about 0.2, keeps constant around 1 for void fractions between 0.2 and 0.9, and

---

* Here "fraction" indicates the ratio of the present value to the nominal value.

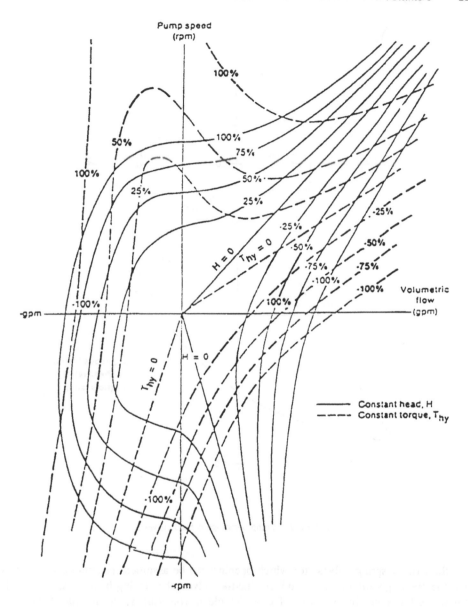

FIGURE 1.   Pump characteristic four-quadrant curves.

finally varies from 1 to 0 for void fractions between 0.9 and 1.0. By adding the hydraulic torque, obtained from the homologous curves, to the friction torque we have the total resisting torque (T). Such a quantity and the moment of inertia of the pump are important because they determine the time required by the pump to stop after the loss of power supply. Regarding the valve component, the importance of simulating its behavior, in certain situations, with a high degree of accuracy, has been already underlined. That is particularly true for safety and relief valves, because of their function in controlling pressure in the hydraulic circuits. Most of the safety and relief valves, used in the plants, are motor or spring valves.

In the case of motor valves, the characteristic curves, giving the flow area and the pressure drop coefficient as functions of the stem position must be provided. Also, if the stem motion is slaved to the variations of a particular physical quantity, stem velocity and time delay between the actuating signal and the actuation itself must be known.

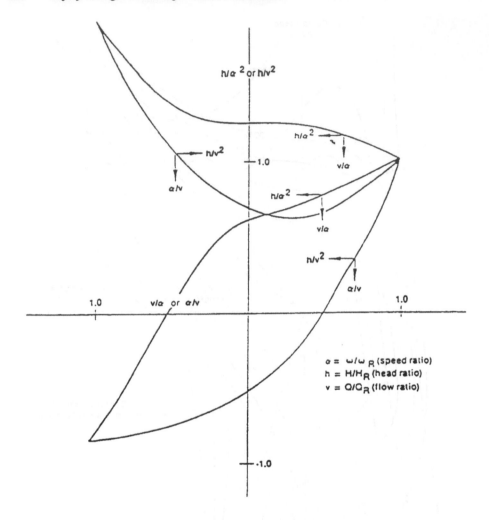

FIGURE 2.   Pump homologous head curves.

In the case of spring valves, for which opening can be assumed instantaneous, the flow area, the opening static pressure, and, for inertial valves, the closing back pressure must be known. Furthermore, in the cases of considerable inertial forces, the inertia of the valve moving parts and the torque provided by the fluid have to be known, since they determine the stem acceleration. Finally, we recall that usually all the valves have very complex grooves, from the point of view of geometry, so the theoretical calculation of the flow rate and the pressure drop across the valve has usually little meaning. For such quantities experimental data that should be provided by the constructor are required.

C. Typical Limits and Problems of Safety Analysis

The final goal of the plant safety analysis is to get a realistic estimate of the safety margins, with respect to initiator events of accidents or transients. Two families (or versions) of computer programs for safety analyses exist: best estimate (BE) programs and evaluation model (EM) programs.

In the first case, computer programs aim at getting the best estimate of reality; therefore,

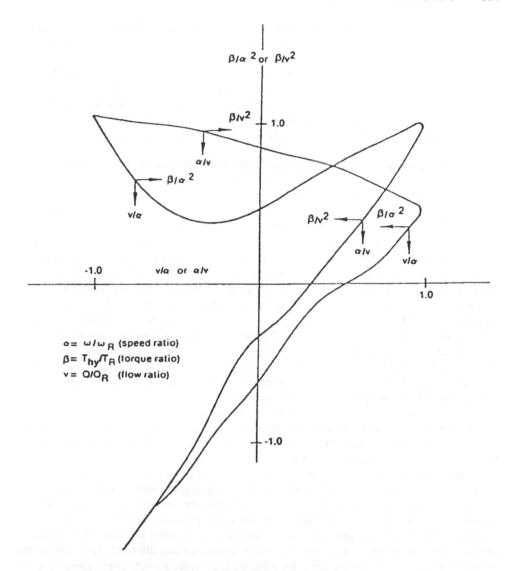

FIGURE 3.   Pump homologous torque curves.

the conservative hypotheses, which are typical of evaluation model programs* aiming at defining the worst hypothetical accident scenarios, are missing.

The construction of nuclear power plants gave a significant impulse to the development of system computer programs, since safety analysis is an essential stage of such plant design. Computer programs now available in the nuclear field are able to simulate most of the phenomena concerning the accident analysis of nuclear power plants.

Restrictions to the improvement process in the simulation of the thermohydraulic behavior of a plant, by means of computer programs, are essentially due to the analysis methods and to the intrinsic limits of the calculating machines (memory capacity, processing time).

---

*   In the nuclear field, the EM computer programs adopt the conservative hypotheses specified by the U.S. Nuclear Regulatory Commission (USNRC) for the Emergency Core Cooling System (ECCS) acceptance criteria. These programs play a different role with respect to the BE ones, which use realistic models in order to analyze the actual plant behavior under accidental conditions.

In this field a quick evolution happened, involving the formulation of thermohydraulic models, able to describe phenomena of increasing complexity, and the development of more sophisticated mathematical methods.

The simplest formulation of the thermohydraulic model of a fluid system considers the three conservation equations (mass, momentum, and energy) for a homogeneous mixture of liquid and vapor in thermodynamic equilibrium conditions, and two empirical correlations describing the generation and removal of energy and evaluating the hydraulic pressure drops. Such a simple model was used in the early computer programs and is still used today for the first-approximation analyses.

This kind of model is clearly insufficient to describe correctly the plant behavior during accidental events, when nonequilibrium conditions exist between phases and reciprocal interactions between phases must be considered (e.g., loss of coolant accidents by rupture of fluid bounding walls). These accidental events are significantly interesting from the point of view of applications. The restrictions of this simple model imposed the development of computer programs of a quite different conception; in such programs the two-phase mixture is treated as two distinct fluids, liquid and vapor, to which the three conservation equations are applied. Moreover, in order to take into account the interactions between phases, three constitutive equations (for the exchange, along the interface, of mass, momentum, and energy), a thermodynamic state equation for each phase (e.g., $P = f(T,v)$), and a state equation for the physical properties (e.g., $c_p = g(P,T)$) for each phase are introduced. Inside these programs, there are also a number of correlations inside the phases, the mass transfer at the interface, the heat transfer between wall and fluid, and the wall friction. Finally, the definition of criteria for the characterization of the flow regimes is provided. In fact, these regimes influence the heat transfer between the two-phase fluid and the structures as well as the entity of transfer phenomena at the interface.[1]

The models adopted in system computer programs to describe the behavior of the two fluids use empirical models, obtained by experimental tests, or semiempirical models suitable to deal with complex phenomena. In both cases, the range of validity and applicability of the models has well-defined limits that the user must know.

Another limit to the accuracy of the computer program, in the prediction of the real behavior of the plant, is connected with the mathematical approximations adopted; theoretical models are approximations of physical processes and the numerical solution is only an approximation of the real one. In fact, thermohydraulic processes are described by systems of partial differential equations which do not have an analytic exact solution. Then numerical analysis methods, adopting the finite differences approximation, are used. They involve the reduction of systematic errors, either by adopting high order models (large memory occupation) or by reducing time and spatial integration steps (high computer time). In the present programs, the influence of the numerical approximations is reduced by adopting solving models of the first order, based on finite differences approximation, and by reducing the time — and space — integration steps. Such an approach has the advantage of providing a more detailed analysis in those transients where thermohydraulic quantities vary very rapidly in space and time domains.

Another kind of restriction is represented by the so-called "methodological restrictions" which lead to two issues: choice of the accidental events and transients to be analyzed (in order to determine the real safety margins of the plant) and influence of the computer program user on the results obtained (choice of the computer program and of the use conditions that are most suitable for studying a given event or transient).

The demonstration of the real safety margins of the plant implies an exhaustive study of all the possible dangerous situations which can occur inside the plant itself. The study of all the possible accidental evolutions is often not feasible, in practice, because direct actions of plant operators cause a great variety of possible scenarios. Due to that, the analysis is often restricted to a meaningful set of events, selected on the ground of appropriate hypotheses about failures in systems and components, occurring simultaneously or not (single, multiple, or consequential failures), as well as on the basis of the kind of event.

Even if we make the assumption that the number of accidental events, selected in this way, is sufficient to show the real safety margins of the plant, it remains impossible to foresee, during the analysis, all the possible evolutions caused by a given initial event, especially in a long-term perspective and because of the human factor, that is widely unforeseeable. Finally, we have to consider the influence of the computer program user on the results obtained. First, a suitable computer program has to be used to study a certain event; the physical models and correlations, implemented in the program, must be able to describe the thermohydraulic phenomena involved. That implies an absolute competence of the program user about the physical models and the correlations used in it, as well as about their capability to simulate real phenomena, their limits, and the ranges of application. Also, computer program users must have adequate experience about the computer response in assessment tests (comparisons between experimental data and program output), and a good sensitiveness in selecting the best-fitted nodalizations to represent the various components of the plant, in function of the transient being analyzed. Training, based on applications of the computer program to experimental facilities, is essential for the achievement of the required competence.

In the field of nuclear energy, several test facilities were built with this purpose, both nuclear and conventional (LOFT, LOBI, SPES).* They allow to verify the real capability of available computer programs to predict the real phenomena, and permit the users to develop a critical skill towards the results obtained; such an approach is indispensable to get a good agreement between the computer program response and the real plant behavior.

IV. INSTRUMENTS FOR SAFETY ANALYSIS

Most of the computer programs, at the present time available, were developed for the safety analyses of nuclear power plants. Therefore, the physical models and the correlations, implemented in them aim at the study of the fluids typical of this kind of plants (light water, heavy water, liquid sodium).

Vice versa, the general organization of the computer programs developed is not strictly connected with the nuclear plants and is usually adaptable to any thermohydraulic system. For example, the programs developed for light water reactors can be (and were) applied, without modifications, to thermoelectric plants and other thermodynamic systems using water as thermal carrier. Moreover, the same programs can be applied to quite different plants, by properly modifying the features related to the physical and chemical characteristics of the thermal carrier fluid (thermodynamic properties, heat transfer correlations, etc.).

FRELAP code is an example; it was derived from a version of the RELAP series, by implementing the physical properties of Freon 12, in order to perform analyses on experimental facilities using this fluid.

In a similar way, a system program as RELAP can be and has been used for the thermohydraulic analysis of chemical plants, oil pipelines, and hydraulic networks in general. A general classification of the computer programs developed for the safety analysis of light water nuclear power plants is shown in Table 1.

Table 2 provides a list of some main computer programs internationally available, with a brief information about the organizations where they were developed, their own characteristics, and the application fields.

* LOFT: E.G. & G., Idaho, LOBI: C.C.R. EURATOM, Ispra, Italy, and SPES: ENEA, Piacenza, Italy.

Table 1
CLASSIFICATION OF COMPUTER PROGRAMS FOR LIGHT
WATER REACTOR NUCLEAR POWER PLANTS

System programs	Component programs	Single phenomena programs
Primary and secondary system	Core	Fuel
		Clad
	Steam generator	General thermohydraulics
		Hydraulic loads
	Pump	
	Vessel	Downcomer
		Hydraulic loads
		General thermohydraulics
Containment	Suppression pool	
Environment		

Table 2
SAFETY ANALYSIS COMPUTER PROGRAMS FOR LIGHT WATER REACTOR
NUCLEAR POWER PLANTS

Computer program	Issuing organization	Program characteristics	Field of application
RELAP 4/MOD7	INEL	Monodimensional analysis; the fluid is considered homogenous and in thermodynamic equilibrium; it uses three conservation equations for the steam-water mixture; several options are provided to allow the user to choose physical models and correlation (e.g., the phase separation and the thermodynamic nonequilibrium models); instrumentation and control systems models are not implemented; punctiform neutronic kinetics model is adopted;	System analysis; primary loop breaks accident analysis for light water reactors; not adequate to simulate operational transients (suitable control system models are missing)
RELAP5/MOD1	INEL	Monodimensional analysis; the fluid is considered non-homogeneous and in thermodynamic nonequilibrium; It uses five conservation equations (two mass and momentum conservation equations — one for each phase — and one energy conservation equation) and two equations for friction and mass transfer at the interface between phases, depending on the flow regime; specific models for choked flow, stratified flow, and concentrated	System analysis; primary loop breaks accident analysis for light water reactors; not adequate to simulate operational transients (the control systems models are not flexible)

Table 2 (continued)
SAFETY ANALYSIS COMPUTER PROGRAMS FOR LIGHT WATER REACTOR NUCLEAR POWER PLANTS

Computer program	Issuing organization	Program characteristics	Field of application
		pressure drops for abrupt area changes are provided; control systems models are implemented; punctiform neutronic kinetics model is adopted	
RETRAN/02	EPRI	Monodimensional, homogeneous, and thermodynamic equilibrium analysis; it uses three conservation equations for the steam-water mixture; it has some options for the selection of physical models and correlations, such as slip between phases, heat transfer in free convection, and and critical flow; control systems models are provided; some secondary system components are described (turbine, steam separator); punctiform and monodimensional neutronic kinetics are available	System analysis; operational transients analysis for light water nuclear power plants; small primary breaks analysis; detailed simulation of control system components
TRAC/MOD1 Different versions for boiling water reactors (suffix B) or pressurized water reactors (suffix P), fast running (suffix F) or "detailed" (suffix D) are available	LANL, INEL	This program is monodimensional for the loops and three-dimensional for the vessel; fluid is considered nonhomogeneous and in nonequilibrium conditions; phase exchanges models dependent on the flow regime are included; models for reflood and rewetting during large break LOCA, stratified flow, and critical flow are implemented; some particular components such as jet pumps and separators are modeled; punctiform neutronic kinetics model is adopted	System analysis; accident and transient analyses in light water reactors (designed for large LOCA analysis); control systems models are not very versatile
BEACON	INEL	Containment program, with multidimensional multifluid thermal nonequilibrium models; it is suited for containment analysis during the first minutes after a large break in a light water nuclear power plant; it can be used to determine the hydraulic loads and jetforces on the compartmental walls	
CONTEMPT	INEL	Containment program, with a lump-nodes model; it is suitable in the analysis of pressure and temperature response in the containment after a break in the primary system of a light water nuclear power plant	
COBRA	INEL	Component program; steady-state and transient thermohydraulic analysis of the core of light water reactors	

Table 2 (continued)
SAFETY ANALYSIS COMPUTER PROGRAMS FOR LIGHT WATER REACTOR NUCLEAR POWER PLANTS

Computer program	Issuing organization	Program characteristics	Field of application
K-TIF	LANL	Component program; two-dimensional thermohydraulic analysis of the downcomer of pressurized water reactors	
K-FIX	LANL	Component program; three-dimensional two-fluids analysis of the break flow in pressurized system	
SOLA-FLX K-FIX-FLX	LANL	Component program; hydraulic loads on vessel internals analysis during fast emptyings	
FRAPCON-2	PNL	Component program; thermomechanical steady-state analysis of uranium dioxide fuel rods	
FRAP-T5 FRAP-T6	INEL	Component program; thermomechanical transient analysis of uranium dioxide fuel rods	
SCDAP	INEL	Component program; fuel behavior analysis during severe accidents in light water reactors	

Note: INEL: Idaho National Engineering Laboratory, EPRI: Electric Power Research Institute, LANL: Los Alamos National Laboratory, and PNL: Pacific Northwest Laboratory.

V. MAIN INTERVENTIONS ON SYSTEM AND COMPONENTS DESIGN

The design with the aim of safety of high-risk industrial plants must be inspired to two objectives: to avoid the occurrence of dangerous situations in the confined systems (preventing action) and to limit the consequences, for the plant and the population, of possible accidents (mitigating action). Prevention is pursued by planning safeguard systems able to limit the excursions of the process critical variables (pressure, temperature, concentrations, etc.) within acceptable values, and by adopting component design and realization criteria tending to increase their reliability.

The mitigation of abnormal situations is pursued by designing systems capable of bringing the plant back to safety conditions. In the following, the main interventions on systems and components design are examined, especially for the systems where pressurization effects are caused by an increase of specific internal energy (due to an unbalance between the thermal power given to the system and the power transferred from the system to the environment).

A. Interventions on System Design to Limit Pressurization

The limitation of the pressurization inside a confined fluid system can be obtained by means of devices, whose action reduces the specific energy of the fluid. Because of the significance of the applications, we shall refer to single fluid systems, in two-phase conditions.

Pressure reduction can be obtained in three different ways:

1. By cooling the fluid contained in the system, by means of suitable components carrying out an effective energy transfer from the system to the environment. This is the case of the steam generator emergency feedwater system in a pressurized-water nuclear power plant. The presence of water, in the shell side of the steam generator, gives the required cooling of the primary coolant and prevents an uncontrolled rise in the fluid specific energy and pressure. Another example is given by the spray systems, acting on the external surface of pressurized liquid fuel tanks (GPL), to prevent unacceptable solar radiation heating.

2.   By injecting a flow of the same fluid, with low specific energy, into the system, thus causing a reduction of the average specific energy. An example is given by the liquid spray system, acting in the vapor region of a steam bubble pressurizer.
3.   By extracting a vapor flow from the system: since the vapor phase of a fluid has a high specific energy, the final result is a decrease of the average specific energy again. Typically, safety and relief valves accomplish this function.

From a functional point of view, systems planned to limit overpressures can be distinguished in active safety systems and passive safety systems. Both systems have a common feature: they intervene when a specified physical quantity reaches a preestablished value (setpoint). Intervention modes are different, however, for the two kinds of systems. For active systems, reaching of setpoint generates a signal that, suitably treated, controls one or more actuators (usually motorized), included in the safeguard system (e.g., power-operated relief valves, PORV). Instead, in the case of passive systems, the physical quantity directly acts on the system, causing its intervention (e.g., safety spring valves).

Some further considerations about the means for limiting overpressures, are necessary.

Cooling the fluid, contained inside a system, by means of a second fluid flowing in a heat transfer component (e.g., the steam generator or the surface of the tank previously cited) is very effective when a good thermal energy transfer between the two liquids is assured. Therefore, in a pressurized-water nuclear power plant, the effectiveness of the energy removal by the emergency feedwater is guaranteed only if the primary loop inventory is sufficient to guarantee the presence of water up to the top of the tube bundle and if natural circulation establishes between the heat source (the nuclear core) and the heat sink (the steam generator).

In the case of a liquid fuel tank, the cooling of a significant part of the tank external surface must be provided (the internal side of such surface must be in contact with the liquid phase).

Regarding the injection of a cold liquid, in order to lower the average specific energy of the system, a consideration has to be made; from the point of view of the final pressure value, the particular position in the system, where the injection is actuated, is unimportant in equilibrium conditions. Instead, the depressurization trend, as a function of time, is quite different whether the injection occurs in the liquid region or in the saturated vapor region. Uniform spraying of cool water into a steam dome of a tank causes a quick depressurization (this effect must be held in consideration in the mechanical design of that component). Locally, in the vapor region, the reduction of the average specific energy is greater than that one corresponding to equilibrium conditions; therefore, the depressurization transient is faster than in the case of injection into the liquid mass (a high heat transfer surface between the cold fluid and the vapor phase must be achieved).

The opening of safety/relief valves, located in the vapor region, is a third means of accomplishing a depressurization. Relief valves are usually motorized valves, with manual or automatic actuation. They are placed in the upper part of the system, where vapor is present, or in the expansion tank, if provided.

Through such valves, fluid with high specific energy is released; consequently, a pressure reduction occurs. If, inside the component on which the valves are located, liquid fluid in saturation conditions exists, some liquid evaporates and a new equilibrium is reached at a lower pressure (vaporization heat is provided by the liquid which, consequently, is cooled).

In some cases, in order to obtain a faster depressurization inside the system, injection of cold water and bleeding of steam, through relief valves, are simultaneously carried out.

Relief valves are an active safety system. An analogous system is the so-called "quick relief" system. It provides a means to remove fluid from the component very quickly, when its pressure increases too rapidly with respect to control device intervention capabilities. Also such a system is of active type.

Instead, safety valves, whose operation is based upon the same principle of relief valves (from the point of view of depressurization), are usually spring valves and, during depressurization transients, act as a passive safety system.

Such valves must lift at a pressure significantly lower than the system design pressure and must guarantee that this limit will never be exceeded, in any situation.

About the operation of safety and relief valves, three issues, having implications in the system design, must be pointed out:

1. These valves are designed to drawn steam from the system; when two-phase low quality mixtures flow throughout them, loss of seal could happen because of erosion during closure. Therefore, in order to avoid an excessive rise in liquid level during depressurization, a sufficiently large diameter of the tank (on which valves are mounted) must be provided.
2. When the released vapor is a dangerous (i.e., chemical) substance or can cause extreme depressurizations, secondary on-off valves should be provided along the relief lines in order to isolate the lines again in emergency situations.
3. Relief capacity of valves is calculated in critical-flow conditions; the pipe downstream the valve must be designed in such a way that back pressure is able to guarantee nominal discharge capacity, in any situation.

Rupture disks provide an alternative passive system, having the same function of safety valves. Such components have some advantages with respect to valves, because, in low pressure range, some restrictions exist in the use of safety valves, that is, due to the presence of moving parts having significant mass with respect to the forces involved. In fact, rupture disks can work with very little pressure differences and, because of the little amount of material used, they are also economically convenient. Furthermore, in the case of fluids easily leading to jamming effects in moving parts, rupture disks are able to maintain the breaking pressure constant and guarantee the intervention. If hazardous fluids are involved, the final breaking of the rupture disks must be considered.

Finally, we want to recall that, if the fluid contained in the system is dangerous for public health, some specific plant solutions for the relief system discharge must be adopted. In such a case, the fluid cannot be released to the environment and suitable storage tank, possibly with steam suppression devices, must be provided. The use of such devices allows to reduce the peak pressure inside the tank, due to the discharge of a given quantity of fluid, or to increase the discharge capacity for a given design pressure of the tank. If the amount of released fluid is expected to be largely variable, depending on the particular kind of transient, an appropriate dimension of the tank is requested, and a safety discharge system (usually rupture disks) must be provided; such a system will release steam into a space which is isolated from the outside. Such a space must be sealed and must be able to contain all the fluid released by the thermohydraulic system, without undergoing excessive pressurizations. In order to limit the volume of this containment some steam damp systems can be provided.

B. Interventions on System Design to Increase its Reliability

One of the essential presuppositions for the capability of the safety systems to accomplish the required functions is the warranty of a sufficient level of reliability and availability. The evaluation of the availability of a system can be performed by using specific methodologies and instruments of analysis, about which we have widely referred to in Section III.

In this section we want simply to recall the importance of carrying out an accurate analysis of reliability, for those systems which — in the safety analysis — prove to carry on an essential role for the safeguard of the plant and for the prevention or an effective mitigation of accidental consequences.

The reliability analysis can put into evidence the existence of critical situations in the system design, which can be reported either to an insufficient degree of reliability of specific components, or — more likely — to operation modes of components and systems which, in certain accidental situations, can prove inadequate.

The reliability study, by giving prominence to the opportunity or the necessity of corrective interventions at the level of system design, can be also used to individuate the possible corrective actions which allow to reach the requested reliability levels.

If the necessary interventions require a substantial modification in the design, it could be necessary to repeat the safety analyses with reference to the new configuration, in order to verify that the introduced modifications do not involve negative variations in the response of the system during the foreseen accidental transients. As an example, interventions of design modifications — either introduced or under plan — for the Italian Standardized Project of PWR Nuclear Power Plant, proposed following the carrying out of a probabilistic risk assessment on such a plant, with reference to events inside the plant itself and aimed at evaluating the annual frequency of damage of the nuclear core, are reported below.

1. Modifications on the suction lines of the pumps of the ECCS. The modifications consist in the doubling of the valves on each of the two lines and the removal of the high pressure recirculation block signal coming from the minimum-flow lines. These modifications produce improvements of the requested recirculation function in the break accidents (loss of coolant accidents, LOCA) or in the transients when, after the failure of the emergency feed water system, the plant recovery is possible with the procedure of "feed and bleed", which requires the high pressure recirculation.
2. Introduction of a device for the test, under operating conditions, of the check valves on the ECCS injection lines. This modification increases the availability of the high-pressure and low-pressure ECCS as well as the rapid depressurization operator action. Also, the high pressure system, required for the addition of borated water, reduces the contribution of a number of initiator events, such as steam generator tube ruptures (SGTR) and secondary breaks.
3. Modifications of the residual heat removal (RHR) system, which consist of the addition of a third valve on each suction line, allowing the test under operation of the check valves and in the modifications of the spare RHR pumps (manually aligned). Both the modifications increase the availability of the system which is required for the long-term cooling after the SGTR accidents. Moreover, the presence of the third valve reduces the possibility of LOCAs from failures of components at the interface between the high- and low-pressure portions of the system.
4. Introduction of a new control system for the relief valves (PORV) of the steam generators. It is based on a new logic which makes the valves to operate in steam dump (prolonged relief) and on the addition of a scram signal in the case of low condenser vacuum. These modifications limit the intervention of the safety valves and, consequently, the probability of the relative missed reclosure, with sensible effects on the mitigation of the turbine trip and loss of feedwater transients.
5. Modification to the loss of fluid test (ESIS) system, for the Boron addition and the reaching of undercriticality within 30 min. This function guarantees the shutdown of the reactor in the case of transients without scram
6. Introduction of the SGOS system. This system is formed by three relief lines — one on each steam generator — which discharge in a tank inside the containment. It allows to avoid the release to the atmosphere of contaminated primary fluid during the SGTR accidents with steam generator fill up. This system has no impact on the scenarios of core damage.
7. Introduction of a spare protection system. The modification consists of the introduction

of an added protection system, with a diversified technology with respect to the integrated protection system (IPS). The project is not yet defined but the functions it will have to accomplish are the following:

- Reactor shutdown function, in the case of IPS failure, required by DISP* criteria.
- Start-up of the ESIS for the cooling of the circulation pumps seals.
- Start up of the two motor driven pumps of the Emergency Feed Water System (EFWS) (these two last modifications are suggested also by the SPS results).

8. Increase of the steam discharge capacity of the relief and safety valves of the pressurizer. This modification is aimed at reducing the pressure peak in the transients without reactor scram.

C. Interventions on Components Design to Increase their Reliability

A cautious choice of the single components plays an essential role in warranting the functionality and the safety of the plant. In the field of nuclear energy and of aeronautics, the potential hazards connected with faulted conditions of relevant components has led to adopt quite sophisticated component design methods. Particular care has to be put for those components which, inside the system, have a specific safety function or, however, are relevant to the safety issue: the reliability analysis results are an essential tool for the definition of the typology of such components, of their functional and constructive characteristics, as well as the operation modes and maintenance.

The classification of a generic component relevant for the safety is performed by categories, on the ground of failure rates, and classes, based on its relative dead time (unavailability).

The first aspect involves the component design criteria, the testing standards, and the maintenance rules; the second one also involves the forms and timings of the verifications and repairs during the operation. At present time, on the ground of the possible malfunctions, components are classified in four categories of increasing reliability:

1. Category 1: normal industrial reliability
2. Category 2: high industrial reliability
3. Category 3: aeronautic industry reliability
4. Category 4: almost absolute reliability

All those normal industrial components, which are not subjected to an effective checking, testing, examination, and maintenance system are classified as category 1. The components classified in such a category must not be the initiator of accidents that can jeopardize plant safety. Category 2 is typical for industrial components of high quality standard. These components are subjected to check, according to the rules issued by the official authorities. Category 3 is typical for components whose quality standard is higher than the highest industrial standard. The safety margins can depend on working conditions less heavy than those ones required by the official authorities and preventive maintenance of a kind usual in aeronautics. Inside this category there are also multiple components consisting of a group of two or three single components, belonging to lesser categories. Such single components are independent, individually checked, and form a unique group (redundant components).

In the category 4 there are those components that could cause very serious consequences, in case of some specific and well-characterized failures. The almost absolute reliability can be obtained by:

* DISP: Direzione Sicurezza Nucleare e Protezione Sanitaria — ENEA, Rome, Italy.

- Variety of redundant components, of lesser category, with the same unique function with respect to the single failure.
- Showing that such failure cannot occur, in specific working conditions, or it results from an evolution of lesser category failures, and also for such evolution some specific conditions, inconsistent with the lesser category failures occurred, are required.

From the point of view of safety, a failure of a component belonging to category 4 must be regarded as "incredible".

For a safeguard system, the lower reliability class of its components, inferred by their relative dead time, is assumed as characteristic reliability class. The purpose of the safety analysis is to get a proper safety degree for each accidental sequence and examine the accidents and the evolution paths of immediate characterization as well as those ones which are less observable. The classification of the components in categories, which are different depending on the required reliability degree or the significance of the function performed, implies that specific building features or design, building, and operating criteria are associated with each category. The adoption of such criteria will lead to some modifications in components, operating modes, checking, and maintenance in order to achieve higher reliability standards where the plant reliability analysis shows such a need.

In the following an example is presented, with reference to nuclear industry.

The goal of safety, for a nuclear power plant, consists in limiting the radiological impact on the environment, as much as possible, in any foreseeable situation. In order to do this, the grouping of the various plant conditions in a number of categories is performed. These categories are characterized in terms of probability, or frequency, of occurrence and radiological consequences. Then, in order to face such plant conditions, suitable systems, able to keep the radioactive releases below fixed levels, must be provided. Such systems have to be extremely reliable, in terms of integrity and/or intervention. Consequently, the systems that form the plant will have a different importance in relation to function performed, characteristics of the treated fluid, and consequences of a possible failure.

Design and realization will have to be carried out with quality standards depending on the importance of the component; this last being defined according to the criteria mentioned above. This is the trend of the regulations in force, which distinguishes four safety classes (SC) for nuclear components and systems: SC-1, SC-2, SC-3, and NNS (nonnuclear safety). These classes are connected with the significance that systems and components have from the point of view of nuclear safety. Such regulations fix* design requirements for each safety class. Particularly, and as an example, for mechanical design of components, the ASME III checking code is used. It fixes more and more strict requisites as the safety class of the component increases. The integral application of this code assures the integrity of pressure confinement, in all the working conditions, both planned and hypothetical, of the component.

REFERENCES

1. **Cumo, M. and Naviglio, A.**, *Thermal Hydraulics*, CRC Press, Boca Raton, FL.
2. **Bird, R. B., Stewart, W. E., and Lightfoot, E. N.**, Fenomeni di Trasporto, CEA, Milano, 1979.
3. **Kirillin, Sycev, and Sejndlin**, *Termodinamica Tecnica*, Editori Riuniti, Rome, 1980.
4. **Viti**, *Lezioni di Idraulica — Meccanica dei Fluidi*, Veschi, Ed., Rome, 1975.

* This is according to 10 CFR 50 Licensing of Production and Utilization Facilities. It states that structures, systems, and components, which are relevant for safety purposes, must be designed, built, set running, and tested with a quality standard depending on the significance of the single systems, from the point of view of the safety function inside the plant.

5. **Cesari,** Circuiti Primari e Tubazioni di Classe 1: Progettazione, Verifiche, Materiali e Normative, ENEA, 1984.
6. **Colombo,** *Manuale dell'Ingegnere,* Hoepli, Ed., Milano, 1985.
7. **Isedi, Ed.,** *Autori Vari Enciclopedia dell'Ingegneria,* Vol. 8, Mondadori, Milano, 1972.
8. **Mainardi,** Esame della Normative ASME per Componenti Nucleari: Documentazione di Recipienti a Pressione, Il Calore n. 2/3, 1975.
9. **Cherubini, Cupini, and Giambuzzi,** Principali Aspetti Fisico-Matematici e Numerici del Retran 02, Parte 1, ENEA Report RTI-TIB-FICS-MATTAPL, (84) 7.
10. **Sinigaglia, Venzi, and Martinelli,** *Meccanica della Frattura,* Clup, Ed., Milano, 1977.
11. **Milella,** Seminari ISMES su La Meccanica della Frattura Bergamo, 1985.
12. **ENEL,** Progetto unificato nucleare, Rapporto Preliminare di Sicurezza, Rome, July, 1982.
13. **Moody,** Dispense del corso di Termodinamica Tenuta al CRE Casaccia, 1985.
14. Safety Evaluation of Westinghouse Topical Reports Dealing with Elimination of Postulated Pipe Breaks in PWR Primary Main Loops, Generic Letter 84-04, U.S. National Regulatory Commission, Washington D.C., 1984.
15. Water Hammer in Nuclear Power Plants, NUREG-0582, July 1979.
16. Evaluation of Water Hammer Occurrence in Nuclear Power Plants, NUREG-0927 Rev. 1, March 1984.
17. **Tong, L. S.,** *Boiling Heat Transfer and Two-Phase Flow,* John Wiley & Sons, New York, 1965.
18. **William, H. and McAdams,** *Heat Transmission,* McGraw-Hill, New York, 1954.
19. **Hetstroni, G.,** *Handbook of Multiphase Systems,* McGraw-Hill, New York, 1982.
20. **Cumo, M.,** Termotecnica Sperimentale, ENEA Serie Trattati, July 1982.
21. **Lahey, R. T., Jr. and Moody, F. J.,** The Thermal Hydraulics of a Boiling Water Nuclear Reactor, American Nuclear Society, 1979.
22. **Marinelli, V.,** Termoidraulica dei Deflussi Bifase nei Reattori ad Acqua Bollente ed in Pressione, CNEN Rep. RT/ING, (75)3, Rome.
23. **Singer, G. L. et al.,** RELAP 4/MOD5 User's Manual INEL Rep. No. ANCR-NUREG 1335, EG & G Idaho, Idaho Falls, September 1976.
24. **Johnson, G. W. et al.,** RELAP 4 /MOD7 (Version 2) User's Manual INEL Rep. No. CDAP-TR-73-036, EG & G Idaho, Idaho Falls, August 1978.
25. Los Alamos Scientific Laboratory, TRAC-PD2, An Advanced Best-Estimate Computer Program for Pressurized Water Reactor LOCA Analysis, NUREG/CR-2054, April 1981.
26. Los Alamos Scientific Laboratory, TRAC-PFI, An Advanced Best-Estimate Computer Program for Pressurized Water Reactor Analysis, to be published.
27. **Ramson, V. H. et al.,** RELAP 5/MOD 1 Code Manual, Vol. 1 and 2, INEL Rep. No. NUREG/CR-1926, EG & G Idaho, Idaho Falls, March 1982.
28. **Moore, K. V. et al.,** RETRAN Computer Code Manual, Vol. 1 to 4, EPRI CCM-5, 1978.
29. **Spore, J. W. et al.,** TRAC-BD1: An Advanced Best Estimate Computer Program for Boiling Water Reactor LOCA Analysis, INEL Rep. No. NUREG/CR-2178, October 1981.
30. **Hargroves, D. W. and Metcalfe, L. J.,** CONTEPT-LT.028:A Computer Program for Predicting Containment Pressure — Temperature Response to a LOCA, INEL Rep. No. NUREG/CR-0255, March 1979.
31. **Niederauer, G. F., Breeding, R. J., and Meier, D. P.,** CONTEMPT-EI.28A A Computer Program for Predicting Containment Pressure-Temperature Transients, EI-81-03, Energy Incorporated Report, February 1981.
32. **Broadue, C. R. et al.,** BEACON/MOD3 User's Manual, INEL Rep. No. NUREG/CR-1148, April 1980.
33. **McFadden, J. H. et al.,** Retran-02 Computer Code Manual, Vol. 1 to 4, EPRI NP-1850-CCMA, November 1984.
34. **George, T. L. et al.,** COBRA-WC: A Version of COBRA for Single-Phase Multi-assembly Thermal Hydraulic Transient Analysis, PNL-3259, July 1980.
35. **Rivard, W. C. and Torrey, M. D.,** K-FIX: A Computer Program for Transient Two-Dimensional, Two-Fluid Flow, Los Alamos Scientific Laboratory Rep. No. LA-NUREG-6623, April 1977.
36. **Berna, G. A. et al.,** FRAPCON-2: A Computer Code for the Calculation of Steady State Thermal Mechanical Behavior of Oxide Fuel Rods, PNL Rep. No. NUREG/CR-1845, January 1981.
37. **Thomas Laats, E.,** Independent Assessment of the Transient Fuel Rod Analysis Code, FRAP-T5, INEL Rep. No. NUREG/CR-1974, April 1981.
38. **Siefken, L. J. et al.,** FRAP-T6: A Computer Code for the Transient Analysis of Oxide Fuel Rods, INEL Rep. No. NUREG/CR-2148, May 1981.
39. **Marino, G. P., Allison, C. M., and Majumdar, D.,** SCDAP: a light water reactor computer code for severe core damage analysis, EGG-M-19082, paper presented at the ANS Int. Meet. Thermal Nuclear Reactor Safety, Chicago, August 1982.
40. **Cumo, M., Farello, G. E., and Ferrari, G.,** Heat Transfer in Condensing Jets of Steam in Water (pressure-suppressure systems), CNEN Rep. No. RT/ING, (77)8.
41. **Richardson, D. C. and Vavrek, K. J.,** Probabilistic Risk Assessment Applications in Nuclear Power Plant Design with Diverse Regulatory Criteria, IAEA-SM-275/37, November 1984.

42. **Ciucci, I.,** Le metodologie di sicurezza dal Corso di introduzione alla sicurezza nucleare e all'analisi degli incidenti, Dic. '86 presso ENEA DISP, Rome.
43. **Valeri, A.,** "Gli studi probabilistici italiani" dal Corso citato, Dic. '86 presso ENEA DISP, Rome.
44. **Valeri, A.,** Modifiche al PUN dal punto di vista dello studio probabilistico di sicurezza, Com. interna ENEA, Gen. 1986.

Chapter 8

CHEMICAL DAMAGE CAUSES — EXPLOSIVE SUBSTANCES MANAGEMENT

Salvatore Ragusa

TABLE OF CONTENTS

I. INTRODUCTION

In industrial activities, some chemical substances, under well-defined circumstances, can produce a high-rate reaction and a sudden pressure development (explosion). Explosions can be also produced by the yielding of a pressurized container. If confined spaces are considered, the main factors influencing explosions and their violence, case by case, have some of the following features:

- Chemical structure of the substance and its particular characteristics (e.g., particle size in the case of a dust)
- Concentration of other substances (e.g., oxidizers, inerting agents, catalysts, inhibitors)
- Temperature, pressure, turbulence
- Ignition sources
- Protection and prevention devices (e.g., vents, explosion suppressors)
- Physical characteristics of the enclosure (e.g., strength)

II. STRUCTURAL CHANGES IN MOLECULES

Changes of chemical structure of the substances by chemical reactions result in an energy-exchange with the outside. The energy absorbed or released when a reaction takes place represents the "heat of reaction".

If energy is absorbed (endothermic reactions) the resulting substances contain more energy than there was in the reacting substances. If energy is released (exothermic reactions) the resulting substances contain less energy than the former ones.

Only transformations developing with energy production may be dangerous. If these transformations are kept under control, the released energy is safely transferred to the environment as thermal energy. The final result is an environmental temperature rise. Usually, heat exchanger is "the environment" in industrial processes. When the substances react at a relatively high rate, the energy released in short time might lead to uncontrolled combustion (fire), to generation of shock waves (explosion), or to both. Therefore, the rate at which the reactions go on is very important from a safety point of view: to maintain the process under control a high-rate reaction would require a reliable relevant heat absorption capability, that is not always accomplishable.

The rate of each reaction is regulated from its own velocity coefficient k. According to the equation of Arrhenius, k is given from:

$$k = Ae^{-E/RT}$$

In this equation, R is the gas constant, T the absolute temperature, EV, the so-called "activation energy", and the constant A are typical of the concerned reaction. We may observe that the rate of a reaction rapidly increases when the temperature increases. The temperature influence is more important at higher values of E.

When the pressure increases, the rate of reaction generally increases too.[1] In some gaseous phase reactions the velocity of a reaction follows the square power of the pressure.

Catalysts are chemicals that, if present even in small quantities, increase the rate of reaction. For that reason, before starting a process the components must be carefully decontaminated to eliminate the undue presence of these agents.[2] Inhibitors (or stabilizers), on the contrary, prevent violent reactions.

III. ADIABATIC DECOMPOSITION TEMPERATURE

Each substance has its own formation enthalpy. It is the energy absorbed or developed in the formation of this substance from the chemical elements. For instance, tables indicate that the formation enthalpy of each mole of carbon dioxide (formed from C and O) is -393 kJ.[3,4] The minus sign remembers that the formation reaction is exothermic. Substances having positive formation enthalpy release this energy when they decompose to chemical elements.

A danger measure of the substances is given by their adiabatic decomposition temperature. That is the temperature reached from the most stable decomposition products (see Section IV) in a constant volume and in adiabatic conditions when the substance decomposes. Once this temperature T_{ad} is known, it will be possible to verify if this temperature is higher than the spontaneous decomposition temperature and so if the spontaneous decomposition can be reached. In this case the situation would be worsened.

Once T_{ad} is known, it will be even possible to calculate the maximum pressure P_f reached in the container according to the law

$$P_f = \frac{n_f T_{ad}}{n_i T_i} P_i$$

Here, n_f is the final number of gas moles and n_i the initial one; T_i and P_i are the initial absolute temperature and the initial pressure.

To obtain the adiabatic decomposition temperature, the enthalpy allowable by the decomposition reaction will be made equal to the enthalpy absorbed from the decomposition products to attain the temperature T_{ad}. The specific heats of all the decomposition products (in the range of temperature concerned) are so needed.

IV. STABLE CONFIGURATIONS

Generally speaking, substances are defined stable when they do not change in the chemical composition even if exposed to air, water, shocks, pressure, or heat. Unstable substances decompose, polymerize, condense, or become self-reactive when exposed to air, water, shocks, pressure, or heat. Violent explosions may result from decomposition of some typical unstable substances such as acetylene, ethylene, ethylene oxide, and hydrazine.

In fact, chemical structures spontaneously tend to move from high energy configurations towards lower energy configurations. Two forces push: the attraction to the state having minimum enthalpy H and the attraction to states characterized by higher disorder, namely to higher entropy S. The two forces constitute one entity, the so-called free Gibbs energy, function of the temperature.[5] Each chemical process, at constant pressure and at absolute temperature T may be analyzed through the free energy variation G from the initial to the final configuration

$$\Delta G = \Delta H - T\Delta S$$

This variation is zero for states in equilibrium, is negative for spontaneous processes and is positive for nonspontaneous processes. When many transformations are possible, the more spontaneous reaction is that going to more stable configuration, namely the configuration characterized from the largest free-energy negative variation. Because this reaction releases the largest amount of energy, just this reaction shall be taken for the safety considerations. Gibbs free energy of various substances can be found in specialized handbooks.[3,4]

Let us now consider a reaction, for instance, a decomposition reaction, the ethylene decomposition. This substance might simply decompose in its basic elements, carbonium and hydrogen, according to the reaction

$$C_2H_4 \rightarrow 2C + 2H_2$$

Table 1
STRUCTURAL GROUPINGS
HAVING POTENTIAL INSTABILITY

Acetylide	Fulminate	Nitro
Amine oxide	N-Haloamine	Nitroso
Azide	Hydroperoxide	Ozonide
Chlorate	Hypohalite	Peracid
Diazo	Nitrate	Perchlorate
Diazonium	Nitrite	Peroxide

From Coffee, R. D., *Safety and Accident Prevention on Chemical Operations*, Fawcett, H. H. and Wood, W. S., Eds., John Wiley & Sons, New York, 1982.

or it may decompose in many other ways. Any mixture of hydrogen, carbonium, or hydrocarbons in fact might be theoretically obtained, the only condition being the respect of the stoichiometric law. One among all theoretically possible reactions will develop more energy than all others do; it should be identified. Then, taking a tentative adiabatic temperature of the unknown reaction products, a computer code should calculate the corresponding Gibbs free-energy variation for each of all the hypothetical reactions and select the more energetic one. For this reaction a mass and enthalpy balance is made to verify if the reaction products reach the adiabatic decomposition temperature chosen in this first tentative. In case of a disagreement, a successive attempt is made, attributing a new adiabatic decomposition temperature to still unknown reaction products. The approximation process goes on until the expectation is confirmed by the mass and enthalpy balance.

V. CRITERIA FOR DETERMINING THE POTENTIAL HAZARDNESS

The structural composition of a substance can give an indication about its intrinsical hazard. The structural groups reported in Table 1 are typical of dangerous substances. Further, during the last years some orientative criteria have been defined to identify the inherent hazard of substances, even if the different criteria have no incontestable value and do not always give coherent indications.

The stability criterion ranks the substances in eight levels,[6] depending on their:

1. Decomposition enthalpy (if more, equal or less than 2.93 kJ/g); the decomposition enthalpy of some particular molecular groups is reported[7]
2. Thermal stability; the substance is slowly heated and checked if an exothermic pic is produced[8]
3. Shock sensitivity; behavior of the substance submitted to (ASME) normalized shocks is observed

The thermodynamic criterion is based on the maximum decomposition enthalpy combined with the difference between the combustion enthalpy and the decomposition enthalpy.[9,10] Three hazard regions are defined: low, medium, and high. According to this criterion, for instance, aniline is classified as a low hazard substance, *p*-nitro aniline as a medium hazard, and picric acid as a high hazard. But the danger presented by the azides is not put in evidence applying this criterion.[6-9]

In the thermodynamic-kinetic criterion,[11] the adiabatic decomposition temperature and the energy activation of the decomposition reaction are taken into account. An empiric index or danger scale is established; at lower levels, the methane is located; at highest levels substances such as vinylacetyl and acetylene are found.

The maximum rate of adiabatic temperature of exothermic decomposition reactions sometimes is taken into account too.[12] Other dangerous characteristics of the materials must be considered such as toxicity, radioactivity, corrosive action, and so on. However, a complete research of the hazard presented from any substance cannot avoid the boundary conditions. Water is harmless when isolated, because of its stability, but it can be very dangerous when, for instance, sodium is present. Hence, a whole safety analysis shall investigate both intrinsic hazards and environmental aspects.

VI. PHYSICAL EXPLOSIONS

Among the more-recurring causes of explosions we find the so-called physical explosions. Their origin can be attributed to existence of physical unbalance. Therefore, we can have vapor explosions, thermal explosions, and boiling-liquid-expanding vapor explosions (BLEVEs).

If one fluid comes into contact with another fluid and the two fluids are at very different temperatures, a violent evaporation of the lower temperature fluid can develop (vapor explosion). Part of the received thermal energy is transferred to the ambient by the violent expansion of fluid at lower temperature. The mechanism of the explosion is not well known. The efficiency (external work/allowable thermal energy ratio) largely depends on many circumstances such as the fluids nature, their temperature, and the way of contact. In some accident reconstructions, efficiencies between 13.4 and 9.4% have been found.[13] Laboratory tests have shown efficiencies of a few percent.[14-16] Assumption of adiabatic expansion can be taken as hypothesis in a safety analysis, because it results in higher efficiencies than can be found in real cases.

Thermal explosions happen in chemical processes in case of unbalance between the heat production of the reaction and the heat removal (by reactor, heat exchanger, pumps). Given a reactive mass in which the concentration of A and B reagent substances is C_A and C_B, the highest reactive mass temperature T allowing a nondivergent reaction is[17,18]

$$T = \frac{E}{R \ln\left[- \dfrac{EAHVC_A^a C_B^b}{ShRT_o^2} \right]}$$

The largest sphere critical radius r allowing a nondivergent reaction is[1]

$$r = - \frac{3RT^2 h}{EAHC_A^a C_B^b} e^{E/RT}$$

and the approximate time t to have an explosion in a thermally isolated reactor is[17]

$$t = - \frac{c_v \rho RT_i^2}{EAHC_A^a C_B^b} e^{E/RT_i}$$

In the preceding equations, specific density of the mass, volume, constant volume specific heat, reaction enthalpy per product mole, heat exchange surface, heat transmission coefficient, mass initial absolute temperature, reactor-cooled wall absolute temperature, gas constant, reaction activation energy, and Arrhenius constant have been respectively indicated with ρ, V, c_v, H, S, h, T_i, T_o, R, E, and A. The sum a + b is the reaction order.

To avoid that the chemical exothermic processes diverge and that an explosion happens, the following preventions must be provided:

- Adequate and reliable heat transfer
- Stirrers more easily transfer the heat outside
- Possibility of quenching (using volatile solutions able to drain out part of the heat of reaction by evaporation,[19] or adding inhibitors)
- Possibility of stopping reactants feeding
- Venting[20]

BLEVEs are typical of liquified gases stored in containers at temperature above their NTP boiling points. These gases remain under pressure (and in the liquid state) until the container keeps its integrity so being close to the atmosphere. But rise of the internal pressure, defects in the container materials, external mechanical causes, and fire exposure (and consequent weakening of the materials strength) may result in failure of the container. At this point, a depressurization wave, having the sound velocity, propagates in the containment. A fast liquid evaporation follows; liquid drops pushed on the containment internal surface causes a real water hammer effect. The expansion process provides further energy for propagation of cracks in the container. The weaker section fails, and the resulting burst spreads around pieces (or missiles) of the container and suddenly mixes vapors and air. If the vapors are combustible, a supplementary risk of fire and explosion is added.

VII. DEFLAGRATIONS AND DETONATIONS

Conditions to have a fire or an explosion are the presence of:

- Combustible substances
- Oxidant substance (e.g., air) mixed with combustible in adequate proportions ranging between two limits, upper and inferior
- Ignition source having sufficient power density to trigger and to maintain the reaction until it will be self-sustaining

Fires generally occur when it is the same reaction to make the mixture of combustible and of oxidant. The burning rate per mixture volume unit is low because of the time needed to reach the adequate mixture composition. For this reason, in fires pressure increases and pressure increase rates are very low. Explosions generally occur when the mixture is already established so that the reaction can proceed at fast rates. If flame travels in the mixture slower than the sound velocity, the explosion will be a deflagration. If the flame travels faster, the explosion will be a detonation. To reach a detonation, important masses of explosive mixtures generally are needed because the flame requires important times (and therefore important volume) to attain the due high velocity. As a consequence, detonations are more destructive than deflagrations.

An equation, the Hugoniot equation, describes all the possible states of a layer of a mixture involved in an explosion. Given the initial pressure p_0, the specific volume v_0, the energy E released per unit mass, and γ, the ratio between constant pressure/constant volume specific heat, by the conservation law Hugoniot had found:

$$\frac{\gamma}{\gamma - 1} (pv - p_0 v_0) - 0.5(p - p_0)(v + v_0) + E = 0$$

The curve (Figure 1) passing through the point p_0, v_0 refers to a layer outside the explosive mixture (E = 0); it represents the possible states of any shock wave. Only the points included between the straight lines $v = v_0$, $p = p_0$, and $v = v_0$, $v = v_0 (\gamma - 1)/(\gamma + 1)$ represent physically possible states. The lower branch of the Hugoniot curve relates to deflagrations;

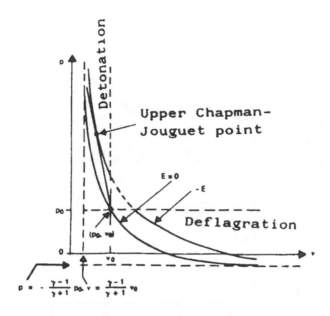

FIGURE 1. Hugoniot curve.

depending on boundaries conditions each point represents a possible state. The upper branch relates to detonations; only one point represents a stable wave;[21] it is called the upper Chapman-Jouguet point. In Reference 22 a method to obtain detonation final pressure, temperature, and wave velocity of explosive mixtures can be found. For instance, in the detonation of the mixture $2H_2 + O_2 + N_2$ final overpressures of more than 17 times the initial pressure, final temperature of about 3400 K, and final velocity of about 2400 m/s are reached.

Not only can combustible gases give explosions but also can dusts of any combustible material. Among explosive dusts are corn, flour, sugar, cocoa, malt, starch, cotton fiber, wood, plastics, urea, carbon, aluminum, magnesium, titanium, uranium, zirconium, and so on. Reference 23 gives a list of more 600 dusts of which explosibility has been proved. Reference 24 also gives a long list of explosive dusts and their explosion characteristics. However, if a dust is not included in these lists, that does not automatically mean the dust cannot explode.

Turbulence makes easier contact between grains and air accelerating the combustion, increasing the probability of having an explosion and increasing the seriousness of the explosion.

Given a dust, its flammability and explosibility increase if the grain dimension is small. In Figure 2 how the dimensions of aluminum grains affect maximum explosion pressure, its maximum increasing rate, minimum energy needed to ignition, minimum explosive concentration, and the so-called "explosibility index" are shown. This index is an arbitrary measure of dusts hazards, related to a standard Pittsburgh coal dust.[25] It is defined as the product of the "ignition sensitivity" (IS) and of the "explosion severity" (ES). The definition of these lasts are

$$IS = \frac{\text{ign temp} \times \text{min energy} \times \text{min conc in Pgh coal dust}}{\text{ign temp} \times \text{min energy} \times \text{min conc in sample dust}}$$

$$ES = \frac{\text{max expl press} \times \text{max rate press rise in sample dust}}{\text{max expl press} \times \text{max rate press rise in Pgh coal dust}}$$

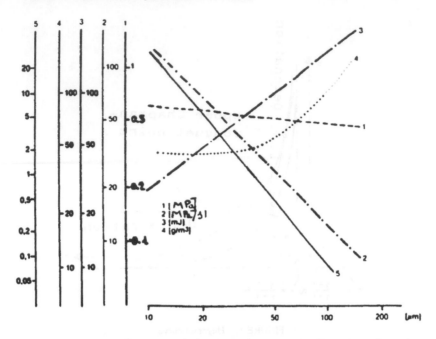

FIGURE 2. Atomized aluminum explosion: 1, maximum pressure; 2, maximum increasing rate; 3, minimum ignition energy; 4, minimum explosive concentration (air); and 5, explosibility index.[56]

Table 2
DUSTS CLASSIFICATION
ACCORDING TO GERMAN
REGULATIONS

Class	K_{St} [MPa m/s]	Explosion
St 1	0 + 20	Weak
St 2	>20 + 30	Strong
St 3	>30	Very strong

Another important parameter for dust classification is the constant K_{St}, useful in vent design. It is linked to the maximum pressure rise and to the volume V:

$$K_{St} = \left(\frac{dp}{dt}\right)_{max} V^{0.333}$$

Dusts are divided in three classes, according to Table 2.

The flame fronts in confined volume explosions have spheric form at the beginning. Then they tend to take the form of the same container. In pipes, the constraint of the wall on the flame amplifies turbulence and hence the combustion surface is enlarged. Therefore, the turbulence region continuously increases in travel along the pipe axis and if the pipe length is more than a critical value, instead of a deflagration, a detonation could occur. As a consequence, the pressure rise due to the explosion will be much more important. In case of deflagration, the final pressure should be of the order of ten times the initial pressure; detonations could raise the initial pressure several hundred times the initial value. On this subject, some research works have been published in recent years.[26-29] Pipe thickness will be calculated to withstand detonation pressures in case pipe length is such to trigger a detonation explosion.

Table 3
PRIMARY MISSILES VELOCITIES

Number of missiles	u	
	Spherical vessels	**Cylindrical vessels**
2	$1.3a^{-0.306}\left[\dfrac{(p_f-p_o)V^{0.653}}{m}\right]$	$1.34a^{-0.608}\left[\dfrac{(p_f-p_o)V^{0.804}}{m}\right]$
10 ÷ 100	$1.3a^{-0.306}\left[\dfrac{(p_f-p_o)V^{0.603}}{m}\right]$	$1.75a^{-0.176}\left[\dfrac{(p_f-p_o)V^{0.588}}{m}\right]$

VIII. MISSILES

Burst shock waves originate missiles and throw them all around. Primary missiles are the fragments of the same bursted containment. Secondary missiles are parts of structures and of process components destroyed in the explosion, tools, and other materials knocked down by the shock wave.

Generally, when a container bursts due to internal detonation, primary missiles have a few grams mass and thousands of meters per second velocities; when a container bursts because of a slow rate internal pressure rise, two missiles having tens of meters per second velocities are generated. In intermediate cases, missiles have some kilograms of mass and hundreds of meters per second velocities. According to experience,[17] it is possible to draw out some formulas to approximately evaluate the velocity of primary missiles. Table 3 shows, for both spherical and cylindrical vessels, primary missile velocities, u, when two missiles or tens of missiles are generated (a = sound velocity in the gas, v = vessel volume, $p_f - p_o$ = relative initial pressure [Pa]).

To complete the scenario, the spatial distribution of the missiles is needed in order to evaluate the probabilities that missiles can strike on important components (from a safety point of view). If hypothesis is made that no preferential direction exists, the probability P that a missile hits the surface dS will be, when $x < u^2/g$:

$$P = \frac{gdS}{\pi^2 xu^2\left(1 - \dfrac{x^2g^2}{u^4}\right)^{0.5}}$$

where g is the gravity acceleration, x the distance from the surface dS and the point of explosion, and u the velocity of the missile. This formula takes into account the gravity force only and does not consider the fluid-dynamic forces (that are important for nonchunk fragments). When $x > u^2/g$ the preceding probability falls to zero; if fluid-dynamic forces are taken into account, there are still probabilities that the dS surface is hit even if $x > 2u^2/g$.[17]

Damages to vessels, piping, and other mechanical components caused from missiles can be grouped in:

- Notching and permanent local deformation which results in weakening the material strength
- Partial penetration; it seems ascertained that rupture will occur as soon as target thickness is reduced to a third of its original value[30]
- Through hole formation with consequent leaks; when the hole dimensions are larger than the critical ones, a brittle rupture will follow

The simplest assumption is to suppose that missile behaves as a punch and that the whole allowable energy is needed to the shear work. Hence a plate will be perforated if:[31]

$$s \leq u \left(\frac{m}{0.6\pi d\tau} \right)^{0.5}$$

where s and τ are the material thickness and shear stress and m, d, and u are the missile mass, diameter, and velocity.

Tests on carbon steel plates (thickness between 3 and 38 mm) hit from austenitic steel cylindrical missiles having diameters between 87 and 160 mm and velocities between 25 and 180 m/s allow to consider that the energy E of a missile able to permanently deform a plate without perforating or cracking approximately is[32]

$$E = 2.9 \times 10^9 \, (sd)^{1.5}$$

In impacts on concrete structures we must consider are

- The scabbing thickness, defined as that barrier thickness which is just enough to prevent scabbing. Scabbing is the peeling of the rear face of the target, opposite to the face of impact.
- The perforation thickness h defined as the barrier thickness that is just enough to prevent perforation.
- The total penetration x. It is the depth that a missile will penetrate into an infinitely thick target or when the target thickness is sufficiently great to prevent rear face scabbing.

To prevent scabbing is an important precaution: concrete fragments originated in the rear face of the concrete target by scabbing might have high kinetic energy and so constitute a real hazard to humans and to process components. Strong and double concrete reinforcement may avoid or limit scabbing.

On the basis of tests, many formulas have been derived. Some of them are shown in Table 4. There σ is the concrete compressive strength and σ_1 is the concrete tensile strength. For Hughes formulas ϕ takes the values 1, 1.12, 1.26, and 1.39 for flat, blunt, spherical, and very sharp nose, respectively. For NDRC formulas ϕ takes the values 0.72, 0.84, 1, and 1.14. In adopting these formulas it is prudent to introduce safety coefficients.[33]

IX. IGNITION SOURCES

Any working area inevitably offers much potential ignition sources. For instance, they can be welding arcs, ovens, hot surfaces, heating systems, tramp iron,[34] and hot spots originated by water hammers hitting on some explosive fluids.[35-36] Everywhere an inflammable or an explosive concentration can be reached, a careful analysis must be done in order to remove any dangerous ignition sources.

Even if it is not ascertained that strong electromagnetic fields can originate sparks able to ignite a flammable or an explosive mixture, theoretical considerations and laboratory tests show that such a possibility cannot be excluded. As a matter of fact any metallic structure in presence of electromagnetic field acts as an antenna. When an accidental or an intentional interruption of the metallic structure continuity happens, a spark could be generated. It seems that any hazard can be excluded if the available power at these metallic structures is less than 1 W.[37-39] When necessary, more usual preventions are electromagnetic shielding and electric bonding.

Table 4
IMPACT OF MISSILES ON CONCRETE

	NDRC[53]	Range of applicability	HUGHES[54]	Range of applicability	CEA-EDF[55]	Range of applicability
Total penetration depth (depth that a missile will penetrate into an infinitely thick target)	$0.0125\dfrac{(\phi m)^{0.5}u^{0.9}}{d^{0.4}\sigma^{0.25}}$ $\left(1+\dfrac{3.86\cdot10^{-5}\phi\,m^{1.6}}{d^{2.8}\sigma^{0.5}}\right)d$	$x \le 2d$ $x > 2d$	$\dfrac{0.19\,\dfrac{\phi m u^2}{d^3\sigma_l}}{1+12.3\ln\!\left(1+\dfrac{0.03\,m u^2}{d^3\sigma_l}\right)}$ —[a]	$40 < \dfrac{m u^2}{d^3\sigma_l} < 3,500$	$\dfrac{0.31\,m^{0.5}u^{0.75}}{d^{0.5}\sigma^{0.375}}$	$2,000 \le \dfrac{m}{dh^2} \le 30,000$ (kg/m³) and $30 \le \sigma \le 50$ (MPa) and reinforcement $150 + 300$ (KG/m³)
Scabbing thickness (thickness needed to prevent scabbing)	$\left(7.91-5.06\dfrac{x}{d}\right)x$ $\left(2.12+1.36\dfrac{x}{d}\right)d$	$\dfrac{x}{d} \le 0.65$ $3 \le \dfrac{s}{d} \le 18$	$\left(1.74\dfrac{x}{d}+2.3\right)d$ $5x$	$40 < \dfrac{m u^2}{d^3\sigma_l} < 3,500$ $t \le 3.5d$ and $\dfrac{m u^2}{d^3\sigma_l} < 40$		
Perforation thickness (maximum thickness that a missile will completely penetrate)	$\left(3.19-0.718\dfrac{x}{d}\right)x$ $\left(1.32+1.24\dfrac{x}{d}\right)d$	$\dfrac{x}{d} \le 1.35$ $3 \le \dfrac{h}{d} \le 18$	$\left(1.58\dfrac{x}{d}+1.4\right)d$ $3.6x$	$40 < \dfrac{m u^2}{d^3\sigma_l} < 3,500$ $t < 3.5$ and $\dfrac{m u^2}{d^3\sigma_l} < 40$		

[a]   Each way reinforcement: front face 0 + 1.5% and back face 0.3 + 1.7%.

However, the more common ignition sources are normally considered to be electrical systems and electrostatic charges.

Prevention of explosive mixture ignition due to electrical systems can be simply made, everywhere possible, locating electrical equipment outside hazardous area. The extent of these areas is defined by specialized codes related to specific dangerous materials involved. Another way is to provide hazardous areas with especially designed electrical equipment. At this subject, National Electrical Code divides hazardous areas in three classes, according to dangerous material present, depending on the probability of hazard, each class is divided into two divisions.

1.   Class I Division 1. Essentially, areas are concerned in which ignitable concentrations of flammable gases or vapors exist frequently or under normal conditions. Electrical equipment used in these areas must be of the explosion-proof type or purged and pressurized type, or intrinsically safe.
2.   Class I Division 2. Essentially, areas are concerned in which volatile flammable liquids or flammable gases are normally confined within closed systems from which they can escape. Generally, electrical components must be in an enclosure approved for Class I Division 2 areas.
3.   Class II Division 1. Essentially, areas are concerned in which ignitible mixtures of combustible dusts may be in suspension. Motors, lighting fixtures, switches, circuit breakers, controllers, and fuses must have dust ignition-proof enclosures approved for Class II areas. Maximum surfaces temperature on equipment must be limited according to dust involved.
4.   Class II Division 2. Essentially, areas are concerned in which ignitible mixtures of combustible dusts may be present as a result of malfunctions. Generally, electrical equipment should be provided with dust-tight enclosures. Maximum surface temperature on equipment must be limited according to dust involved.
5.   Class III Division 1. Essentially, areas are concerned in which easily ignitible fibers or materials producing combustible flyings are processed. Generally, the electrical equipment must be the same as for Class II Division 2 areas.
6.   Class III Division 2. Essentially, areas are concerned in which easily ignitible fibers are stored. Generally, the electrical equipment does not appreciably differs from that used in Class III Division 1.

The more specific types of equipment used in explosion hazard areas are explosion-proof apparatuses, purged and pressurized apparatuses, and intrinsically safe equipment.

Explosion-proof apparatus is designed according to three basic criteria: it must withstand internal explosion of explosive mixtures, it must prevent propagation of internal explosion to the surrounding flammable atmosphere, and its external surface temperature must not be high enough to ignite the surrounding explosive mixtures. It is so recognized that explosive mixtures may enter the enclosure and that there is the possibility of ignition within the enclosure.

Purged- and pressurized-type apparatuses are supplied with positive pressure source of clean air or of inert gas. Controls prevent energizing the container until ventilation have been established and provide deenergizing when the supply is interrupted.

Intrinsically safe equipment and wiring is so designed that even in abnormal conditions the electrical energy released is not enough to provoke ignition of a specific hazardous mixture.

In common technical language, static electricity is defined as charges of the same sign confined in an isolate region of the space, leaving out the consideration if the formation process has been static or dynamic. Usually, static electricity may appear when relative motion of substances is produced, one or both of these substances frequently being a poor conductor of electricity. More common examples of static electricity producers are flow of fluids in pipes, motion of pulverized materials, belts in motion, gases flowing through an opening, and moving vehicles.

When a body, the human body included, become electrically charged, a difference of potential V is established between the body and any other surface as in any electrical condenser. When the electrical field existing between the two surfaces of this condenser, having the capacity C, exceeds the dielectrical rigidity of the interposed air (30,000 V/cm), a discharge occurs. In the discharge, the potential energy of the condenser is transformed in thermal energy. This energy, $CV^2/2$ or $Q^2/2C$ where Q is the electric charge, can be strong enough to trigger explosion in an explosive mixture. In optimal conditions, ignition energy of some dusts can be as low as 0.3 mJ;[23] a minimum ignition energy of about 0.2 mJ have many hydrocarbons,[22] while for hydrogen it falls even to 0.03 mJ or still to less.[23,24] It is not difficult to find static charges of the order of 10^{-7} C in some dust processes such as transfer, screw feeding, and crushing.[40] In this case, capacities of about 5 picofarad (very usual in industrial activities) could rise to stored energies of 1 mJ. A statically charged human body put into contact with the earth could discharge energies as important as 50 mJ.

The best prevention, when practical, is to remove the potentially flammable or explosive substance from locations were electrostatic charges can be present. Of course, this prevention is not possible when the same flammable or explosive substance (e.g., as gasoline flowing in pipes) produces electrostatic charges. A general means to reduce charge production is to slow the fluid or dust velocity. The diminution of charges production linearly (or even more than linearly) depends on velocity.[41,42] Use of conductivity additives in liquids prevents the buildup of hazardous potentials.

Any other applicable method able to increase conductivity of surface, to drain static charges, is useful in preventing charges accumulation. Humidification of atmosphere is one of these methods. A level of approximately 70% relative humidity should be reached. However, this way may be practiced only where the surface on which charges accumulate belong to materials, as paper or woods, that reach an equilibrium with the atmosphere and at the condition that these surfaces are not excessively heated.

Electrical connections of metal parts (bonding) to earth create preferential paths to the charges that in such a way leave the surface on which they try to accumulate. Ground rod resistance of some hundreds of thousands of ohms are sufficient; current to be handled is very small. For these reasons a ground connection which is adequate for power circuit protection will be certainly adequate in our case too.

Ionization of air is pursued to make it conductive and to remove charges from paper, fabrics, and driving belts. This ionization can be obtained with radioactive isotopes (α and β emitters). The presence of radioisotopes implies a potential radiological risk.

Another way of ionizing air is reached by installation of bars equipped with a series of needle points bonded to earth. When the electrical field created in the points by the charges becomes higher than dielectric rigidity of air, discharges are triggered. The arch is then quenched as soon as the charge falls below a certain level; hence this is a discontinuous process. A continuous air ionization can be obtained applying on the above needle points an high voltage alternating tension. However, this method presents an intrinsic hazard due to the presence of high tension and so asks for specific precautions.[43,44]

Finally, a particular charge formation process, the lightning, asks for an adequate protection. The lightning discharge must be intercepted before it strikes on a potentially flammable or explosive atmosphere. The only way to do that is to discharge the lightning current harmlessly to earth. A system of conductive musts and overhead wires electrically connected to the earth (Faraday cage is the borderline case) should envelop the area to be protected. Steel tanks in direct contact with the ground can be considered inherently grounded; pipes

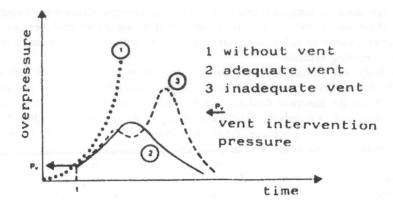

FIGURE 3.   Explosions in confined volumes: overpressures (qualitative).

entering a tank must be metallically connected to the tank itself at the point of entrance; all joints must be riveted, bolted, or welted to assure a good electrical continuity.

X. VENTS

Whenever possible, hazardous operations and related equipment should be placed outdoor or in segregate protected buildings. Otherwise, at least, they should be placed in one-store building or in the top story of a multistory building. Besides, being nevertheless possible, indeed, that a combustion explosion, or a runaway explosion, or any other kind of explosion can occur despite adopted preventions, the accident consequences must be limited as much as possible by venting.

A deflagration combustion developed in a vessel can increase the initial pressure by a factor of 7 ÷ 10. A deflagration combustion in a pipe or in a duct can develop into a detonation due to the turbulence facilitated by the presence of a long confined space; as a consequence the initial pressure can increase by a factor as important as some hundred times. When practicable, any container should be designed in such a way to withstand the resulting maximum stress even in the occurrence of an explosion. The safety factor could be not too higher than 1 if the final pressure and the material answer are exactly known. Certainly it is not the case in piping explosions, due the final pressure indeterminateness to ascribe to lacking of recognized theories and to incompleteness of experimental works.

Buildings cannot be designed to withstand explosions because it would be technically and economically impracticable. Therefore, the only possible way to protect the buildings is to vent them.

Venting (safety valves) is usual protection for vessels and for other process components.

The most effective vent for releasing an explosion pressure would be an open vent, without any cap or diaphragm. Only very few processes and kinds of containers can allow such a type of vent, almost all cases asking for vent closures. Vents must be designed to open quickly and automatically under increasing pressure of the protected containment and in such a way to obstruct the venting flow as low as possible. Moreover, the delay in opening must be reduced because the rise in pressure will be increased from any delay. Delays in opening vents are caused by inertia of the same vent closure and by the overpressure required to open the vent. Finally, the open surface of vent must be as large as possible to limit the final pressure in the container. As qualitatively shown in Figure 3, inadequate vent area will result in an undesidered pressure peak.

Basic rough law for vents is that vent area ratio of two vessels equals the two-thirds power of their volume ratio.[45] It is not prudent to apply this rule in extrapolating from a larger vessel to find the vent area of a smaller one.

Table 5
RUNES EQUATION.
VALUES OF C FOR
SOME SUBSTANCES

Substance	$C\left[Pa^{0.5}\right]$
Organic dusts	6.8
Ethylene	10.5
Hydrogen	17

From Cross, J. and Farrer, D., *Dusts Explosions*, Plenum Press, New York, 1982.

Table 6
VESSELS MAXIMUM OVERPRESSURE $\left[MPa\right]$ AS A
FUNCTION OF THE VENT/SURFACE RATIO

Maximum overpressure	$\dfrac{\text{maximum dimension}}{\text{minimum dimension}} \leq 3$	$\dfrac{\text{maximum dimension}}{\text{minimum dimension}} > 3$
$P > 0.1$	$0.24(\dfrac{S}{AE})^{0.7}$	$0.24(\dfrac{4S}{A'E})^{0.7}$
$P \leq 0.1$	$1.23(\dfrac{S}{AE})^2$	$1.23(\dfrac{4S}{A'E})^2$

From Singh, J., *Chem. Eng.*, 103, 1979.

Runes Equation

$$A_v = \frac{Cl_1l_2}{P^{0.5}}$$

is applied in determining building vent area,[46] with the limitation $l_3 \leq 3 (l_1 l_2)^{0.5}$. In the preceding formula, P is the overpressure that can be held by the weakest point of the building, $l_1 l_2$, and l_3 are the three building dimensions (l_3 being the longest one) and C [Pa $^{0.5}$] a dimensional constant which depends on flame velocity and flame temperature. This constant can be found in Table 5 for some materials.

For venting of gas containers, the expressions of Table 6 can be applied. They have been derived from a set of tests and supposing that the safety valve is destroyed as soon as its intervention pressure has been reached. Values of S can be found in Table 7. A is vent surface/vessel internal surface ratio. A′ is vent surface/vessel minor cross-section ratio. E is the efflux surface/vent surface ratio, normally taken equal to 0.6.

For dust explosions, venting nomographs as in Figure 4 are largely used.[47,48] Given the container volume V, the limit pressure P, the dust class St_1, St_2, and St_3, the vent area A_v can be found. Each nomograph refers to a vent pressure intervention P_v. These nomographs cannot be employed when vessel length/vessel minimum dimension ratio is more than five and when the vessel volume is larger than 1000 m³.

Vents of dangerous substances should be conveyed towards safe locations, i.e., confined or unconfined volumes that can accept the release. For any appreciable duct length, the duct cross-section surface should be twice or more that of the vent device. It should be taken into account that due to the friction in the duct the maximum pressure reached in the vessel

Table 7
VALUES OF S FOR SOME
GAS-AIR STOICHIOMETRIC
MIXTURES IN SPHERICAL
VESSELS; INITIAL
CONDITIONS 0.1 MPa, 298K

Gas	S
H_2	0.044
CH_4	0.0085
C_2H_2	0.032
C_2H_4	0.014
C_3H_8	0.0096

From Bradley, D. and Mitcheson, A., *Combust. Flame*, 237, 1978.

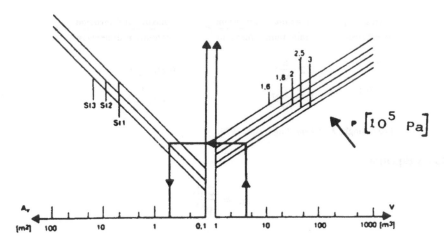

FIGURE 4.   Dust explosions: vent area nomograph (intervention pressure 0.15 MPa).

will be higher than in the conveyor absence.[23] In case of gas burning in the conveyor, turbulence developed in the duct could lead to transition of the deflagration into detonation.

XI. INERTING AND SUPPRESSING SYSTEMS

An inert gas present in a burning mixture warms at the expense of combustion products; the flame temperature is lowered and the combustion becomes more difficult. If inert gases are present in adequate proportions, they are so a sound means to prevent fires and explosion in enclosed spaces. In dusts, when the amount of inerting gases present is not sufficient to prevent an explosion, a reduction of the maximum rate of pressure rise is obtained. This reduction is linearly decreasing with the increasing of the inert gas percentage.[49]

The most common inertizing gases are nitrogen and carbon dioxide. Argon, helium, chlorinated-fluorinated hydrocarbons, or even combustion products are used. It should be verified that no incompatibility exists between the employed inert gas and the substances or the materials of the process.

Many publications of the U.S. Bureau of Mines furnish for different gases, vapors, and dusts the maximum permissible oxygen percentage to prevent ignition. Quantities in excess of that indicated as the minimum needed are suitable.

Inert gases can be introduced in two ways: continuously or completely filling fixed volumes. In both cases, attention will be paid to obtain and to maintain a complete diffusion of the inert gases in the volume to inertize in such a way that combustible mixture pockets are avoided.

Sometimes combustible gases are used as inert gas. It can be the case in combustible storage tanks. Their vapor space can be filled, for instance, with natural gas to maintain a fuel-rich (oxygen-deficient) atmosphere, above the flammable limits.

Once an explosion has happened, to have its development hindered, stopped, and to have its effects limited, suppression systems are installed.

These systems consist of a detection device and of a suppressing device. They are fast-acting because the allowable time for intervention is too short (of the order of milliseconds); any delay increases the consequences of the incipient explosion, among them the stresses of the containment to be protected.

Generally, suppressors are scarcely effective or quite ineffective when the pressure rise rate is very high (as in hydrogen or in aluminum explosions).

The more commonly used detectors are sensitive to pressure rise, to pressure rise rate, to visible light, or to radiation emitted from the incipient explosion.

The suppressing agent is a liquid or a dust compatible with the substances and with the process material. Normally, pressurized suppressors contain the suppressing agent under nitrogen pressure; an explosive charge opens the container and the extinguisher is released. Ejectors must be able to diffuse everywhere the suppressing agent in sufficient quantities. In dust explosion an orientative figure of the needed dust extinguisher is given by $7 \, V^{0.67}$ kg, where V [m^3] is the volume to be protected.[50]

REFERENCES

1. **Ragusa, S.**, Introduzione all'analisi del rischio nell'industria, in S.I. Safety Improvement, c. Europa 12 Milano, 1986, chap. 3.1.2.
2. **O'Brien, G. J., Gordon, M. D., Hensler, C. J., and Marcali, K.**, Thermal stability hazards analysis, in *Chemical Engineering Progress*, VAICRE, New York, 1982, 46.
3. **Dean, J. A.**, *Lange's Handbook of Chemistry*, McGraw-Hill, New York, 1979.
4. **Weast, R. C., Selby, S. M., and Hodgman, C. H.**, Eds., *Handbook of Chemistry and Physics*, 62nd ed., CRC Press, Boca Raton, FL, 1982.
5. **Metz, C. R.**, *Physical Chemistry*, McGraw-Hill, New York, 1976.
6. **Coffee, R. D.**, Chemical stability, in *Safety and Accident Prevention on Chemical Operations*, Fawcett, H. H. and Wood, W. S., Eds., John Wiley & Sons, New York, 1982.
7. **Smith, D. W.**, Runaway reactions and thermal explosion, *Chem. Eng.*, 79, 1982.
8. **ASME**, Standard Method for Assessing the Thermal Stability of Chemicals by Methods of Differential thermal analysis, E 537-76, 1982.
9. **Davis, E. J. and Ake, J. A.**, Equilibrium thermochemistry computer programs as predictors of energy hazard potential, *Loss Prev.*, 1973.
10. **Treweek, D. N., Claydon, C. R., and Seaton, W. H.**, Appraising energy hazard potentials, *Loss Prev.*, 21, 1973.
11. **Stull, D. R.**, Linking thermodynamics and kinetics to predict real chemical hazards, *Loss Prev.*, 67, 1973.
12. **De Haven, E. S.**, Using kinetics to evaluate reactivity hazards, *Loss Prev.*, 1979.
13. **Anderson, R. P. and Armstrong, D. R.**, Comparison between vapor explosion models and recent experimental results, in AIChE Symposium Series 138 on Heat-Transfer Research and Design, 1974, 31.
14. **Buxton, L. D., Nelson, L. S., and Benedick, W. B.**, Steam explosion triggering and efficiency studies, in Fourth CSNI Specialist Meeting on Fuel-Coolant Interaction in Nuclear Reactor Safety, FC 14/P17, Bournemonth, England, 1979, 387.
15. **Fauske, H. K.**, On the mechanism of uranium dioxide-sodium explosive interactions, *Nucl. Sci. Eng.*, 95, 1973.

16. **Kottowski, H. M., Mehr, K., and Grossi, G.,** Vapour explosion studies in a constrained geometry and forced fragmentation and mixing, *Am. Soc. Mech. Eng. Heat Transfer Div.*, 17, 1981.
17. **Baker, W. E., Cox, P. A., Westine, P. S., Kulesz, J. J., and Strehlow, R. A.,** A Short Course on Explosion Hazards Evaluation, Southwest Research Institute San Antonio, Houston, 1978.
18. **Semenov, N. N.,** *Some Problems of Chemical Kinetics and Reactivity*, Pergamon Press, Oxford, 1959.
19. **Austin, G. T.,** Hazards of commercial chemical operations, in *Safety and Accident Prevention in Chemical Operations*, Fawcett, H. H. and Wood, W. S., Eds., John Wiley & Sons, New York, 1982.
20. **Huff, J. E.,** Computer simulation of polymerizer pressure relief, *Loss Prev.*, 45, 1973.
21. **Paterson, S.,** Detonation and Other High Temperature Phenomena at High Pressures. Introduction, Discussion on the Faraday Society, 1956, 22.
22. **Lewis, B. and Von Elbe, G.,** *Combustion, Flames and Explosion of Gases*, Academic Press, New York, 1961.
23. **Field, P.,** *Dust Explosion*, Elsevier, Amsterdam, 1982.
24. **NFPA,** *Fire Protection Handbook*, 15th ed., National Fire Protection Agency, Quincy, MA, 1985.
25. **Jacobson, M., Cooper, A. R., and Nagy, J.,** *Explosibility of Metal Powders*, RI 6516, USDI, U.S. Bureau of Mines, 1964.
26. **Bollinger, L. E., Fong, M. C., and Edse, R.,** Experimental measurements and theoretical analysis of detonation induction distances, *Am. Rocket Soc. J.*, 588, 1961.
27. **Flessner, M. F. and Bjorklund, R. A.,** Control of gas detonation in pipes, *Loss Prev.*, 113, 1981.
28. **Sargent, H. B.,** How to design a hazard-free system, *Chem. Eng.*, 250, 1957.
29. **Sutherland, M. E. and Wegert, H. W.,** An acetylene decomposition incident, *Loss Prev.*, 99, 1973.
30. **Lannoy, A. and Gobert, T.,** Evaluation of hazards from industrial activities near nuclear power plants — Deterministic and probabilistic studies, in 5th Int. Conf. Structural Mechanics in Reactor Technology, Berlin, 1979.
31. **White, E. A.,** Missiles: penetration capability, Appendix 6, in Canvey An investigation of potential hazards from operations in the Canvey Island/Turroch area, HMSO, 1978.
32. **Ohte, S., Yoshizawa, H., Chiba, N., and Shida, S.,** The strength of steel plates subjected to missile impact, in *6th Int. Conf. Structural Mechanics in Reactor Technology*, North-Holland, Paris, 1981.
33. **Romander, C. M. and Sliter, G. E.,** Model test of turbine missile impact on reinforced concrete, *Nucl. Eng. Des.*, 331, 1984.
34. **Merwin, R. F.,** Explosions caused by tramp iron reduced by approved magnetic separators installations, in Int. Symp. Grain Elevator explosions, PB-292728, U.S. Department of Commerce, 1978.
35. **Faeth, G. M. and White, D. F.,** Ignition of hydraulic fluids by rapid compression, *Am. Soc. Naval Eng. J.*, 467, 1961.
36. **Stull, D. R.,** Fundamentals of fire and explosion, *Am. Inst. Chem. Eng.*, Vol. 73, 1977.
37. **DIN 57848 Teil 3,** Gefaehrdung durch elektromagnetische felder explosionsschutz, Entwurf, 1983.
38. **Maddocks, A. J. and Jackson, G. A.,** Measurements of radio frequency voltage and power induced in structures on the St. Fergus gas terminals, in *The Radio and Electronic Engineer*, Vol. 51, 1981, 187.
39. **Rosenfeld, J. L. J., Strachan, D. C., Tromans, P. S., and Searson, P. A.,** Experiments on the incendivity of radio-frequency breakflash discharges (1,8-21 MHz c.w.), in *The Radio and Electronic Engineer*, Vol. 51, 1981, 175.
40. **Gibson, N.,** Decomposition, fire and explosion hazards from dusts: a review, in *Inst. Chem. Eng. Symp.*, Series 3, Rugby, England, 1981.
41. **Boschung, P. and Glor, M.,** Methods for investigating the electrostatic behaviour of powders, *J. Electrost.*, 205, 1980.
42. **Lamouche, R.,** Etude de l'élimination de l'électricité statique sur des hydrocarbures, in *Mesure et Élimination de l'Électricité Statique Nuisible*, Challande, R., Ed., Eyrolles, Paris, 1973.
43. **Cross, J. and Farrer, D.,** *Dusts Explosions*, Plenum Press, New York, 1982.
44. **Henry, P.,** Manifestations et effets des charges electrostatiques, in *Mesure et Élimination de l'Électricité Statique Nuisible*, Challande, R., Ed., Eyrolles, Paris, 1973.
45. **Donat, C.,** Pressure relief as used in explosion protection, *Loss Prev.*, 87, 1977.
46. **Runes, E.,** Explosion venting, *Loss Prev.*, 63, 1973.
47. **NFPA 68,** Guide for Explosion Venting.
48. **VDI RICHTLINIE 3673,** Druckenstlastung von Staubexplosionen, VDI-Verlag Dusseldorf, 1977.
49. **Nagy, J., Cooper, A. R., and Stupar, G. M.,** Pressure development in laboratory dust explosions, RI 6561, U.S. Bureau of Mines, 1964.
50. **Bartknecht, W.,** Gas, vapor and dust explosion. Fundamentals, prevention, control, in Int. Symp. Grain Elevator Explosions, U.S. Department of Commerce, PB-292728, 1978.
51. **Singh, J.,** Sizing vents for gas explosions, *Chem. Eng.*, 103, 1979.
52. **Bradley, D. and Mitcheson, A.,** The venting of gaseous explosions in spherical vessels II — Theory and experiment, *Combust. Flame*, 237, 1978.

53. **Kennedy, R. P.**, A review of procedures for the analysis and design of concrete structures to resist missile impact effects, *Nucl. Eng. Des.*, 183, 1976.
54. **Hughes, G.**, Hard missile impact on reinforced concrete, *Nucl. Eng. Des.*, 23, 1984.
55. **Berriaud, C., Sokolovsky, A., Gueraud, R., Dulac, J., and Labrot, R.**, Comportement local des enceintes en beton sous l'impact d'un projectile rigide, Nucl. Eng. Des., 457, 1978.
56. Explosibility of Metal Powders, R.I. 6516, U.S. Department of Interior, Bureau of Mines, 1964.

Chapter 9

THE ROLE OF INSTRUMENTATION AND CONTROL

Fausto Zambardi

TABLE OF CONTENTS

I. INTRODUCTION

Physical and chemical processes in industrial plants where highly dangerous substances or large amounts of energy are managed represent main sources or risk.

A large variety of instrumentation and control (I and C) devices is presently available and is extensively used to maintain industrial processes under controlled conditions.

Nevertheless various occurrences may happen such as design errors, process disturbances, equipment failures, and operator error, which can cause degradation from normal to abnormal and accident conditions and danger or damage for the public health and for the environment can arise.

Due to the risk of social, health, economic, and environmental consequences, a safety demand must be accomodated in preventing accidents and mitigating their effects, and a systematic approach for safe design and operation of industrial plants is needed.

The global performance of plants in both normal and degraded conditions relies on control and protective actions actuated either automatically or manually by means of I and C systems. These actions are to be planned and verified on the basis of a safety evaluation in which plant scenarios are to be defined and analyzed with reference to the behavior of the process variables for any configuration of plant systems and equipment with given degradation assumptions; the effects due to the performance and the malfunctions of the instrumentation in itself must also be taken into account.

The safety aspects concerned with I and C functions and the design considerations applicable to I and C in high-risk plants are discussed afterwards.

A. The Role of Instrumentation and Control in Industrial Plants

Control, protection, and supervision are the main functions required to instrumentation systems in a plant. These systems are constituted by an interconnection of different I and C devices and equipment distributed throughout the plant and consisting basically of the following: sensing and measuring devices, processing equipment, and manual commands and information displays.

Sensing and measuring devices (sensors and transducers) are devoted to the generation of electrical signals representative of physical variables to be monitored. In some cases pneumatic signals are also used. Variables to be monitored include process- and equipment-related parameters (pressure, temperature, level, flow, displacement, speed, voltage, electric current, vibration, etc.) and environment-related parameters (temperature, humidity, radiation, concentration).

Monitoring in high-risk plants implies generally an extended measuring range for process instruments to cover the large changes induced in variables by abnormal and accident conditions; additional instrumentation can be required to detect these conditions and to alert about the environment status. Monitoring of environment conditions is referred to the measurements of variables such as temperature, humidity, radiation, toxic, or flammable substance concentration in plant areas where presence of personnel or of critical equipment is foreseen; generally environ monitors have the capability of providing area alarms when preset values are surpassed.

Special care may be reserved to the detection of the release and the diffusion of noxious substances outside the plant. In all identified release points, flow, chemical composition, and/or radiological activity level are monitored.

Diffusion is controlled by monitoring the meteorological status (wind speed and direction) and the concentration of released substances.

Processing equipment is aimed at the elaboration of information coming from the sensors (i.e., from the process and associated equipment and from the environment) and to the generation of signals directly usable for control, protection, and supervision purposes.

Signal processing begins with signal conditioning (filtering and buffering, current to voltage or analogic to digital conversion, multiplexing); the conditioned signals are then treated with instrumentation modules performing functions related to the specific application (function generators, trip units, logical gates, lead/lag units, etc.). Computers are also employed in lieu of hardwired systems. Processing equipment is often collected in centralized locations (e.g., the control room) which can be better protected from environmental effects due to abnormal plant conditions.

Information displays and manual commands are devoted to the supervision and control actions of the operating personnel; they include indicators, recorders, alarms, computer terminals, status lights, switches, push-buttons, and manipulators; this equipment is located on control room panels or in other centralized locations and interfaces with processing equipment.

1. Control

Control function of instrumentation is referred to the generation of command signals devoted to maintain plant process and equipment parameters at the values or under the laws of variation required for normal plant operation. The control of process involves the establishment of on/off and analogical control actions. In both cases, the actual process parameters are compared with the required values (setpoints) and a signal is generated for switching on/off or for regulation. Analogical control actions include proportional, integral, and derivative actions. Standard controller modules are in use with provisions for manual control and indication to the operator.

Control of active equipment is related to the control of their status (operating or not, mode of operation in correlation with the status of interfacing equipment) and requires logical processing of status variables (by means of "or", "and", and "not" logic operations, etc.).

Control functions are often realized by means of process computers because of their inherent versatility in performing any kind of elaboration and in accomodating all regulations and controls required for plant operation with sufficient speed. Operator control and supervision can also be optimized by means of computers.

In high-risk plants the control function is important as a mean to maintain potentially dangerous processes in a safe condition for the workers, the environment, and the public.

Generally, process parameters can be optimized with respect to risk factors, such as the undesired generation of noxious substances, and an effort has to be made to avoid the process from deviating from the design condition; in this case a control system with high characteristics of precision and reliability can contribute substantially to the risk reduction.

2. Protection

Protection function of instrumentation is referred to the generation of command signals (protection systems) for the actuation of those protective actions which are anticipated to be necessary when the plant goes through abnormal or accident situations and control systems are no longer able to manage the plant.

The purpose of the protective actions is to reestablish safe conditions timely and with an adequate margin against safety limits. For this reason I and C components of protection system must have low response time to assure maximum rapidity in the intervention and high precision with respect to the point of intervention to assure the effectiveness of the protective action.

Built-in instrumentation is provided for manual command and for indication to the operator.

Protective actions may have a preventive or mitigative function. Prevention is referred to limit the possibility of accidents while mitigation relates to minimize accident consequences. In both cases the problem is, by an instrumentation point of view, to monitor the effective plant status, to perform an elaboration of logical variables representative of the status with the objective of discovering any deviation from normal conditions or any degradation, and to generate the appropriate command signals for equipment which must actuate the protective functions. The equipment performing the logical elaboration of the actuation signals can be based on solid-state devices or electromechanical relays.

Recently protection systems have been developed based on computers and microprocessor networks. On principle control systems used to regulate the operation of equipment actuating the protective functions are part of the protection system. The importance of protection systems and their criticality for the safety is, in some way, dependent of the type of plant as it is outlined below.

In fossil-fueled power plants a challenge to the public health arises routinely from the continuous release to the atmosphere of toxic substances generated during the combustion, because of the impossibility of complete filtering. The additional concern from accidents can be considered low, so that the need for a protection system is dictated more by the necessity of assuring, for economical reasons, the integrity of the machinery. What is needed for public safety in this case, by an instrumentation point of view, is mostly an optimization in control and supervision systems to limit the plant releases and to control their diffusion in the environment.

In nuclear power plants dangerous radioactive substances accumulate in the nuclear fuel as a result of the fission process. An accidental loss of integrity of fuel and of other barriers (reactor coolant pressure boundary, containment) can cause a large and sudden release with fatal consequences for the environment and the public. In this case it is of vital importance to have the availability of a powerful protection system which timely gives rise to the appropriate prevention and mitigation actions.

3. Supervision

Supervision function of instrumentation consists in providing the operator with all available information about the plant operating status and the environment, this information being useful for the manual command of control and protection systems and for the assessment of plant and environment status.

This assessment can constitute (as in the case of nuclear plants) the ultimate basis for taking extraordinary emergency measures in order to safeguard the public health.

Information is collected from monitoring and processing systems and is supplied to the operator by means of dedicated devices such as indicators, status lights, alarms, recorders, cathode ray tubes, and computer printouts. Most of this equipment is concentrated in centralized control areas (main control room and other control centers of the plant).

Devices for supervision and alarms of environmental conditions (radiation level, dangerous chemical substance concentration) are distributed through the plant (and in its proximity).

The control of chemical properties of process fluids is performed by means of sampling stations.

Supervision is in close correlation with plant operation procedures. The employment of supervision in the management of accident situations is aimed at the recognition of plant events and of appropriate actuation of safety actions. Based on a more elastic utilization of supervision systems, a tendency is in nuclear power plants to pass from event-oriented procedures to symptom-oriented procedures; as a result a minor diagnostic effort is expected from the operator.

B. Safety Relevance and Defense in Depth for Instrumentation and Control

As previously outlined, a first characterization can be made for I and C according to the degree of implication in safety functions. This characterization should be aimed at the definition of most appropriate requirements to apply from design to operation and could be carried to the point of identifying a classification in terms of safety relevance and engineering requirements (as it is in the nuclear field).

In judging the extension of the I and C set to consider, a comprehensive approach can be kept based on a defense in depth concept. The most immediate level of defense is represented by control and supervision systems devoted to normal plant operation; an appropriate quality level for these systems is an excellent guarantee of not going out of the normal condition and of containing the risk.

At the next level can be placed the instrumentation for preventive actions and for mitigative actions; the most stringent engineering requirements must be applied for a significant risk reduction.

The safety level depends on the integrated performance of protection systems specifically devoted to safety and of nonsafety-related control systems; operator contribution also must be taken into account as it is affected by supervision systems performance. An ultimate verification of I and C effectiveness should bring to the exclusion of:

1. Nonsafety-related control system failures involving transients more severe than those which can be successfully faced by protection system by design, or affecting adversely any operator programmed action
2. Nonsafety-related control system failures which cause, with high frequency, the deviation from normal plant conditions, the violation of normal condition limits (for plants in which operation is subjected to technical prescriptions), or the intervention of safety systems

Such an investigation raises many questions to discuss.

Let us consider, as an example, a chemical reactor equipped with safety valves in which an endogenous process takes place for the generation of innocuous substances; the verification in this case should address the possibility of control malfunctions involving;

1. The chemical process changes from endogenous to exogenous.
2. The rise in temperature changes the chemical process giving place to the generation of noxious substances.
3. The rise in pressure causes the opening of the safety valves.
4. The noxious substances are released through the safety valves with consequences for the environment and for the public.

A possible conclusion could be a design revision looking for increased precision and reliability in control systems and a different protective scheme.

II. DESIGN OF INSTRUMENTATION FOR HIGH-RISK PLANTS

A. Plant Operability Considerations

The main goal to pursue with the instrumentation and control is the operability of the plant for all conditions, including accidents, and the homogeneity of the technical choices related to I and C.

The operability is linked to the supervision and control functions demanded to the operator and performed by means of instrumentation systems. The incidence of human error and the availability aspects of the instrumentation itself must be addressed. A careful consideration must be reserved to the man-machine interfacing problems and to the management of instrumentation (with particular emphasis on maintenance aspects).

Another factor affecting the operability is the degree of correlation between the frequency and nature of human actions on instrumentation and its functional/physical allocation in the plant. This correlation dictates exigencies of centralization/decentralization of instrumentation, leading generally to the definition of the following plant areas.

1. Process area or "field" which is the general plant area where a sharing of instrumentation (sensors, transducers, etc.) must be provided and its physical interfacing with process/equipment/environs must be assured as required by the monitoring functions. Operator intervention is generally for test and maintenance.
2. Instrumentation room, which is a particular area of the plant devoted to the allocation of processing equipment.
3. Control room and other centralized control positions, which are devoted to the allocation of control and supervision instrumentation mostly involved in the plant conduction by the operator.

Management of instrumentation for plant operability requires a suitable degree of uniformity in technical characteristics in view of establishing a simple interface with plant personnel.

It is convenient that this uniformity be implemented at each phase, as appropriate, from conceptual design to maintenance procedures. It is also deemed that the homogeneity of instrumentation (or the good interfacing between different types of instrumentation) can really constitute an effective basis for training programs.

An example of instrumentation design affecting directly plant operation is in the following.

Digital input signals can be generated either by field instrument switches or by instrumentation channels including analogue transmitters in field and trip units in the control room; in the second case an improved plant reliability can be achieved because of a decrease in field instrument drift and reduction in calibration time and personnel presence in potentially dangerous areas for maintenance.

In more general terms, the extension of automation affects directly the plant operability. In fact there are many industrial processes which can be practiced only by employing a large amount of automatic control means instead of direct operator intervention. The reasons for restricted human actions include the following:

1. The location in which the process takes place is inaccessible because of the environment degradation involved by the process itself.
2. The conduction of the process requires a sophisticated elaboration of the pertinent informations.

3. Control action effectiveness is strictly dependent on rapidity and precision of interventions.
4. The criticality of certain control actions (in particular those referred to safety) requires a very low probability of lacking interventions.
5. The safety status assessment requires the knowledge of process and environment parameters.

In front of the mentioned conditions the automation is given the role of:

1. Allowing the remote control of inaccessible processes
2. Performing all required elaboration, assuring with a proper level of precision, rapidity, and reliability all the required control actions
3. Supplying the operator with all information for the status assessment.

The quality degree of automation is essential for assuring the required performance; the safety instrumentation must have a quality level higher than common industrial quality. Further, the role of instrumentation must be specified in front of the basic plant conditions, as delineated in the next paragraph.

B. Design Basis for Instrumentation Important to Safety

The detailed specification of I and C functions and performance requires the identification of a coordinated set of design basis which should include at least the following information:

1. Complete identification of the processes taking place in the plant and of the associated equipment for their control (valves, pumps, heaters, etc.).
2. Physical parameters to be monitored, steady-state valves, extreme range of variation, maximum rate of change, and number and location of sensors for variables with spatial dependence.
3. Disturbances and component malfunctions affecting the process performance.
4. Control action required for the process (manual, automatic, response time, accuracy, etc.).
5. Design events (large equipment failures and process degradations) for which safety functions must be provided and corresponding protective actions performed by the equipment actuation: manual or automatic initiation of protective actions (depending on the dynamic of the design event).
6. Operational value (control system setpoint) and safety limit for each variable.
7. Margin between safety limits and protection setpoints (values of parameters at which a protective action must be initiated); this margin must accomodate instrument inaccuracy and drift, transient overshoot, and response time.
8. Parameters to be displayed, signaled, or alarmed to the operator for his supervision function and to allow him to perform the appropriate control and protection actions when needed.
9. Auxiliary supporting equipment (electrical power supply, compressed air supply, means of control of environmental conditions, etc.).
10. Maximum range and rate of change in power supply (voltage, current, frequency) and environmental conditions (temperature, humidity, radiation, etc.). Number of abnormal occurrences anticipated in the plant life.
11. Conditions that can compromise the functional integrity of the safety equipment (operator error, spurious operation of safety equipment protective device, spurious actuation of fire suppression system, fire, flood, missiles, earthquake, aircrash, sabotage).
12. Reliability analysis methods applicable to I and C components as well as to other

components of safety systems performing protective actions, to demonstrate that global reliability goals are achieved (refer to Section 2, Chapter 4).

C. Design Aspects

Aspects to be considered in the design of instrumentation devoted to safety purposes are identified for the following areas:

1. Number of sensors
2. Location of sensors
3. Transducer, location, and connection
4. Input signal
5. Manual commands and automatic control
6. Interactions between protection and control
7. Operating and maintenance bypass
8. Test and calibration
9. Management of instrumentation
10. Fault tolerance
11. Independence and separation
12. Qualification
13. Electromagnetic interference immunity
14. Component classification selection and quality
15. Software design
16. Information for accident supervision
17. Auxiliary equipment
18. Essential auxiliary equipment

1. Number of Sensors for a Single Variable

More than one sensor can be required because of:

1. Redundancy exigencies imposed by availability and reliabilities considerations
2. Spatial dependence of the variable to be measured
3. Functional exigencies; if a single sensor cannot cover the variable range or if more than one function is associated to the same variable (e.g., the loss of a sensor shall not compromise both automatic and manual control)

2. Location of Sensors and Measuring Points

The location may affect the performance, for instance, a flow sensor must be placed far enough from pipe bends, the position of pressure-sensing taps on a pipe depends on the physical status of the process fluid (liquid, gas, vapor).

3. Transducers, Location, and Connection with the Process

Sensing lines connecting transducers to the process must have a slope (according to the type of process fluid) and must be equipped with root valves (for isolation from the process), venting and draining points, isolation valves or restricting orifices (for loss of fluid limitation in case of downstream rupture), and heat-tracing equipment (to prevent the lines from freezing if exposed to low temperatures).

Structural characteristics of sensing lines must be consistent with those of process tanks and pipes (acting as pressure boundary or barrier to process fluid release). The association of transducers to the same sensing line must be functionally verified in front of the common failure resulting from the line fault.

Allocation of transducers directly on the process equipment should be limited; mounting on local panels is preferred because a protection against environmental effects is provided as well as a centralization for maintenance operations.

It is convenient that panels be located baricentrically with respect to measuring points and appropriately accessible by personnel.

Auxiliary equipment of panels must include instrument valving for test, bypass, flushing, and drainage; power supply and internal lighting must also be provided as a minimum.

Appropriate identification of all components by tag or color coding is required.

Instruments and their enclosures are so designed to prevent water or powder entry. In presence of flammable or corrosive atmosphere, additional precautions must be applied as follows:

1. Adoption of explosion-proof enclosures
2. Adoption of a purging system (an inert gas is continuously supplied to the instrument environment)
3. Proper design of circuits to prevent electrical faults from being cause of undesired ignition
4. Relocation of instruments in safe areas
5. Adoption of pneumatic instrumentation

4. Input Signals

When different functions are performed for the same process variable, different and dedicated sensors must be used to generate the required input signals so that the loss of a single sensor will not result in the loss of all functions.

Analogue transmitters and trip unit channel sets are preferred (because of the improved reliability with respect to field instrument switches) for the generation of input digital signals for relevant functions.

Inputs for protection functions must be derived, to the extent possible and practical, from direct measures of the plant variables specified in the design basis; derivation from other equipment such as indicators and recorders is not allowed.

5. Manual Commands and Automatic Control

Commands must be available in the control room for the manual initiation of the automatic protective actions at the division level. A division is a complete and independent set of instrumentation actuation and auxiliary equipment which can perform a given safety function, more than one division is normally made available for the same safety function. If manual control is provided at component level, this does not have to defeat the command at division level. A special manual command (override) can be provided to defeat an automatic signal; this feature can be applied in the case of automatic initiation of redundant divisions (in accident situations) and the operator, at his own judgment, decides to go on with the safety action with only the minimum sufficient equipment, switching off (i.e., overriding) all other equipment which had been automatically activated. Overriding is also applied to activate a system interlocked by the logic, once its integrity has been verified.

In principle overriding involves a challenge to plant design basis, therefore it should be allowed only if a safety increase is deterministically shown to result.

Displays are required to indicate the status of the protection system (available or rendered inoperative) and the actions performed. Displaying needs to be on a division basis accordingly with the architecture of the protection system.

Manual or automatic control is possible for regulation functions supported by automatic/manual control stations; the transfer from automatic to manual can be "bumpless", that is, an automatic tracking feature keeps the manual signal at the same level of the automatic control signal (setpoint) so that shocks on the processes are avoided in the transfer. The amount of operator manipulations for manual commands has to be minimized, as well as the involved equipment. The treatment of commands (manual or automatic) must permit the completion of the safety action once this has been initiated.

6. Interaction Between Protection and Control

Protection systems do not have to be functionally degraded because of adverse actions of control system (and more generally, nonsafety systems) resulting in a condition which requires by itself protective actions.

An obvious case in which protection unavailability would result from control misoperation could be if the same variable were processed with common equipment for protection and control. For instance, in a temperature-controlled process in which the temperature sensor fails low, the difference signal with respect to the regulation setpoint goes high and asks for an indefinite increasing of temperature (which leads to the process degradation and accident if no limitation intrinsic to the process happens), at the same time the protection setpoint cannot be exceeded by the failed low temperature signal and no protection signal can be generated for the actuation of safety action.

Effective means to avoid interactions between protection and control are the functional diversification and/or the physical diversification.

Functional diversification means that control and protection are based on different physical variables (all of them effective for the process status representation) which are processed by dedicated equipment so that a failure will be confined and will not involve the degradation of both protection and control functions.

Physical diversification means that diverse equipment is employed as an effective barrier against common cause failures to process the variable for control and for protection; diversity in equipment is in terms of diverse principle of operation or diverse manufacturing.

The ultimate objective to achieve for inhibiting all possibilities of dangerous interactions should be the functional independence between protection and control.

7. Operating Bypass and Maintenance Bypass

Different modes of plant operation can need an automatic or manual bypass of safety functions. If the permissive conditions for the activation of the operating bypass are not met, this activation must be automatically prevented.

An automatic restoration action must also take place if the operating bypass is no longer allowed because of changes in plant operation mode.

The maintenance bypass is different from an operating bypass; it consists in the removal from service of a part of I and C for test, calibration, or maintenance whose safety function must be assured by other unbypassed equipment.

8. Test and Calibration

Provisions for testing and calibration must allow the verification of safety functions performance during plant operation and without any loss of functional availability of safety equipment. The frequency of this verification has to be in accordance with the reliability goals of the safety system. Sensor checks must be included in this verification and can be performed by cross-checking between different channels, by perturbing the monitored variable, or by applying to the sensor a test input instead of the process variable. In the last case the provisions for operating bypass must apply.

9. Management of Instrumentation

Maintainabilty must be assured by means of design provisions such as autodiagnostic and modularity to facilitate the recognition and replacement of faulty units. Instrumentation must be physically identified as being safety equipment. Controlled access by the personnel is required for any action to minimize the possibility of undue operations.

10. Fault Tolerance

Instrumentation must be designed such that functions can be performed also in the presence of faults. A typical way to achieve fault tolerance is by redundancy, that is, to perform the required function by means of a few identical and complete sets of equipment; in presence of a failure (such as open circuit, short, short to ground) affecting a complete set of I and C (division), the function will be actuated by other unaffected sets.

Attention must be devoted to the loss of redundancy due to maintenance bypasses; the bypassed division must be put in an intervention state or the remaining divisions must satisfy the fault-tolerance requirement.

Common cause failures in redundant divisions must be excluded, otherwise diversification provisions must be applied.

Another way to achieve fault tolerance is to apply the fail-to-safe concept; for instance, a relay which must be deenergized to actuate a safety function can be considered fail-safe with respect to the electrical supply; if this fails, the function will remain assured by the consequential deenergization.

11. Independence and Separation

Redundant safety divisions of I and C must be functionally independent and physically separated between them and from other instrumentation not safety related. A way to achieve independence of safety instrumentation with regard to the interconnected instrumentation is by electrical isolation (interposition of optical couplers, fuses, isolation amplifiers, or transformers).

Physical separation is to maintain instrumentation integrity against various causes of degradation due to equipment in proximity. These causes include the following:

1. Missiles generated by the rupture of high-pressure tanks and pipes or heavy rotating equipment, or by the explosion of high voltage terminal box
2. Pipe whip resulting from high energy process lines
3. Jet impingement following high pressure pipe rupture
4. Fire
5. Flood (external and internal)

Physical separation can be achieved by means of adequate separation distance or by segregation or by interposing physical barriers.

The implementation of independence and separation is necessary to satisfy the fault-tolerance requirements for redundant I and C systems.

12. Qualification

Qualification is the demonstration that the functional and performance requirements of instrumentation are met for the environmental and interface conditions resulting from all events which are anticipated to affect the plant for its entire design life.

The demonstration can be conducted by analysis, by test, or by a combination of both; operating experience data can also be used. A unique demonstration is sufficient to support the qualification of a family of components, provided that these components can be considered functionally and physically identical or similar; if it is conducted by test care must taken for assuring the representativity of the prototype component being tested with respect to the family of components being qualified. Because of its uniqueness the demonstration can give only assurance about functional and performance aspects; any result in terms of reliability figure is excluded.

The functional requirements depend on the single instrument and its particular use for safety purpose. For instance, a trip unit must pass from nontripped to tripped state, a recorder must record, and a contact must open and close.

The performance requirements consist in the way the functions are required to be performed, i.e., accuracy, response time, input/output impedance, gain, etc.

The environmental conditions refer to all those factors which are cause of degradation for the equipment being qualified and include temperature, humidity, corrosive agents, radiation, vibration, powder, etc. The entity of these factors must be established by means of an analysis of the environmental effects consequential to the anticipated plant conditions, from normal conditions to accidents. External events such as earthquake and aircrash give place to vibratory stimulus for components and must also be taken into account, as well as electromagnetic disturbances due to lightning.

A special degradation factor to be considered is the aging which is linked to the time in itself and to the time-growing effects of other degradation factors.

The interface conditions are linked to the anticipated changes in performance of functionally interfaced equipment such as voltage and frequency of power supply.

A main goal of qualification is the preservation of redundant instrumentation against common cause failures and the determination of the qualified life, i.e., the duration time for which the component demonstrates satisfactory functioning. The qualified life is often less than the plant design life so that it constitutes an important entry condition for replacement programs.

13. Electromagnetic Interference Immunity

Instrumentation is for its nature susceptible to electromagnetic interferences (EMI) affecting the plant environment as a consequence of storm lightning, electrical power circuit operation, and intrasystem and intersystem disturbance effects.

To assure the integrity of critical I and C, also by a functional point of view, a large variety of provisions is applicable: shielding of peripheral walls of plant building and of critical instrumentation room, connection of common reference signal and instrumentation ground/shielding at a dedicated point of the ground electrical network, coaxial/triaxial cable, cable shielding and twisting, current signal transmission on long distances, and filtering.

14. Component Classification, Selection, and Quality

To facilitate the choice of instrumentation the relative importance of functional requirements and of technological aspects of instrumentation in itself must be evidenced and a proper classification must be introduced to allow the determination of congruent and graduated sets of engineering requirements.

Factors calling for higher class and more stringent requirements attribution include:

- Direct involvement with safety functions
- Exposition to critical interface conditions
- Constraints to physical accessibility
- Criticality of component failure consequences in front of system design characteristics
- Inherent complexity of instrumentation in itself
- Employment of nonconsolidated technologies of realization
- Lack of operating experience in the specific application
- Procurement difficulties

The implementation of the engineering requirements must be enforced by means of a quality assurance program whose severity must also be commensurate to the factors mentioned above.

Instrumentation components modules and systems must be subjected, from design to operation and maintenance, to quality assurance requirements which enforce standards, rules, and practices for high quality.

The quality goal is in terms of minimization of maintenance requirements and failure rate. The reliability goal for instrumentation system needs to be verified with analytical methods.

15. Software Design

The software for computerized safety systems should satisfy a "free error" condition. The limitation of design techniques does not allow to obtain deterministically a free error software as soon as its complexity increases. The technique for producing good software has arrived to a maturity point.

Main aspects of this technique are the top-down approach (systematic derivation of detailed specification from more general specifications with a step-by-step method of development and modularization), structured programming (minimization of instruction types and avoidance of unconditioned jumps), and the verification and validation (V and V); verification is the step-by-step process of verifying the various design phases by means of analytical methods and tests; validation is the concluding demonstration of software correctness based on a comparison, mainly conducted by test, between the initial functional requirements and the final product performance).

16. Information for Accident Supervision

Plant and environment variables and status must be supplied to the operator for the assessment of accident conditions; a classification of this information is needed because of its criticality and the information channels, from sensor to display, and must address design requirements graduated according to the criticality level.

Variables providing information for safety assessment include, in a decreasing order of criticality, those necessary for the manual control of safety functions for which no automatic control exists, those referred to the accomplishment of safety functions and to the structural integrity of physical barriers, those related to the operating status of safety systems, and those referred to environment contamination.

Design requirements applicable to instrumentation channels include the following:

1. Measuring range commensurated to values of variables which are allowed by physical laws (i.e., they are deemed to go well beyond the nominal values)
2. Qualification of channels components (from sensor to display)
3. Validation of information (it is needed a level of redundancy or diversification in the channels such that a malfunction is not cause of ambiguous or erroneous information)
4. Independence and separation of redundant channels
5. Power supply from standby noninterruptible sources
6. Channel availability constraints for the plant operation
7. Quality assurance program from design to maintenance
8. Dedicated and continuous display and recording
9. Overlapping of instrument spans if a few instruments are used to cover the variable range
10. Identification of supervision channels
11. Control of access to equipment
12. Direct derivation of information
13. Human factors

17. Auxiliary Equipment

Equipment not supporting directly the safety functions (e.g., testing equipment) is required not to degrade the safety function because of its presence.

18. Essential Auxiliary Equipment

Auxiliary equipment required to support the safety functions of instrumentation is to be considered as an active part of the safety system. An example is the electrical power supply.

The architecture of the supply system must reflect the redundancy characteristics of the supplied instrumentation and divisional assignment exigences must be respected. The design considerations exposed before (fault tolerance, independence, separation, qualification, etc.) must be taken into account, as applicable.

Availability, capability, and uninterruptibility are to be assured by means of standby power sources including motor/generator sets, inverters, and back-up batteries.

Coordinated, electrical protection is required in the distribution system to prevent the loss of supply to all instrumentation loads because of individual electrical faults (e.g., the short circuit at a single load). Supply performance (voltage, frequency) is not to be degraded by transients deriving from switching on/off of heavier electrical loads.

Another example is the air supply system for pneumatic instrumentation. Besides the system requirements which are similar to those outlined for electrical supply systems, care must be taken with respect to quality of air which must be oil and contaminant free and with minimum size particle content.

III. INSTRUMENTATION FOR SUDDEN IDENTIFICATION OF DEGRADATIONS

In high-risk plants the opportunity arises to identify precociously the degradation affecting the components as a result of operational stresses and aging. This identification, on one side allows extraordinary maintenance or replacement actions to be undertaken in addition to those preventively programmed (conditioned maintenance), on the other side, and consequently permits to anticipate and prevent unusual interventions of control systems and shocking effects from protection system actions.

The role played by instrumentation in the systems for sudden identification of degradations is outlined in the examples of applications described afterwards.

A. Structural Integrity of Passive Equipment Continuous Monitoring

Passive equipment, such as vessel and pipe, acting as pressure barriers with respect to the contained fluids, must guarantee a high level of confidence in maintaining their structural integrity because of the consequences in case of sudden ruptures. The same happens for structural members, restraints, and heavy machinery. As these ruptures are preceded by an incubation period of time with growing degradation, it is worthy to monitor the equipment for discovering any premonitory symptom.

Monitoring can be performed either on a continuous basis by means of on-line dedicated instrumentation channels, or periodically, in the form of in-service inspection (ISI), by qualified personnel with movable devices.

Physical variables of interest for the on-line monitoring of structural integrity include acoustic emission associated to fluid leak and crack growing in the structural material, strain of the material, and relative displacement between mechanical components.

Measuring devices include displacement sensors and load cells (for loading forces); these sensors are based on capacitive, inductive, piezoelectric, or reluctive effect. A kind of sensor used extensively is the strain gage. Strain is measured by means of resistive strain sensor consisting essentially of a conductor (or semiconductor) with small cross-sectional area applied on the surface of the material being monitored. A deformation of the sensing element follows by the stress-induced deformation in the material and results in a change in resistance and a bridge output voltage variation. Acoustic emission is measured with piezoelectric sensors. A preamplifier provides for the raising of the signal level and the low frequency noise associated with operation of equipent is filtered out, various operations can be performed on the output signal (from simple recording to sophisticated correlation analysis).

B. In-Service Inspection

ISI permits deeper knowledge about the conditions of equipment in a plant; attention is particularly focused on critical areas such as the verification of integrity of welds and restraints.

Instrumentation is a fundamental factor to implement certain diagnostic methods employed in ISI, mainly those which are based on ultrasonic and eddy current techniques.

Equipment for ultrasonic technique perform basically the following functions:

1. Supply the ultrasonic excitation energy to the part under control by means of an emitter transducer
2. Pick-up the response by means of a receiver transducer (which can be the same emitter); the response is correlated to the integrity of the material
3. Amplify and process the signal for presentation of information

Transducers are based on piezoelectric effects. Eddy current technique employs transmitting and receiving units to introduce an electrical current into the material under control and to observe the interactions between the material and the impressed current.

Instrumentation is also important to control equipment used for other diagnostic methods such as those based on radiographic and magnetic techniques.

Diagnostic methods utilized for ISI require generally a sophisticated elaboration to interpret the information returned by the controlled parts and to draw the knowledge about the integrity status of the material; for this aspect the most effective approach is to rely on automatic data-processing devices.

Because of the accessibility problems arising for certain plant areas (high temperature, radiation, toxicity, etc.), a tendency develops through the remotization of the control of diagnostic equipment employed in ISI and through robotics applications; the introduction of fully automated diagnostic means would permit, on the other hand, to have a deep knowledge of equipment integrity status on a continuous basis.

C. Loose Parts Monitoring System

In process lines and tanks where a continuous circulation of fluids is established, a challenge to material integrity can arise because of solid parts dispersed in the fluid. These parts can be caused, for instance, by the detachment of internal mechanical components such as bolts, or can consist of extraneous objects (poor maintenance). These parts are transported by the flow and impact on the internal surface of process equipment with a kinetic energy that can be significant.

Systems to monitor loose parts, as applied in nuclear power plants, are based on the detection of the vibrations induced by impacts by means of piezoelectric sensors. These sensors are placed in locations such as bends of pipe or lower area of tanks where impacts of loose parts are more likely to occur; parts that weigh 0.5 kg is a possible design goal.

The elaboration of the sensor signal includes generally a level rise by preamplification and a filtering of extraneous noise. The resulting signal puts the impacts in evidence in the form of spikes whose amplitude is related to the energy of the impact. Time correlation between the corresponding spikes indicated by the monitoring channels in different points addresses the exact location of impacting parts.

The performance of the system is subject to a number of spurious signals due to operational causes, e.g., the end of travel of valve stems, therefore, the effectiveness of the system depends on the level of knowledge about all sources of noise in the process equipment.

D. Active Component Monitoring

Critical active components (rotating equipment, valves) can be continuously monitored during the operation to ascertain that their functional performance is maintained unchanged with the time. A typical variable to take under control in the case of electrical motors is the internal temperature; in this regard a number of temperature sensors may be distributed in the windings and high temperature alarms are made available.

Other variables of interest are the pressure in the lubrification circuits and the temperature in cooling circuits. Instrumentation channels include, besides sensors, indicators and alarms.

The mechanical performance of active component is generally based on the detection of vibrations. Piezoelectric sensors are normally used for detection. The monitored variables are then treated for indication, alarm, and recording. A way to utilize records is to compare periodically the present vibration signatures with reference signatures; this allows to put in evidence degradations for bearings, rotating shafts, or control valves.

E. Leak Detection

An evaluation of the integrity state of plant equipment can be obtained by monitoring the presence of the process fluid in the environment, or its effects.

A first step consists in monitoring the controlled ways of loss, e.g., the drain lines from glands, by means of flow and temperature detectors. A second step is to monitor the liquid leak collecting points; the level can be directly indicated or more simply the frequency of starts of drain pumps actuated by the level switch intervention. A third step is the measure of the atmosphere chemical composition to detect the presence of substances released by the process because of the degradation in equipment. Continuous monitoring channels sensing the atmosphere concentration can be employed or laboratory analysis can be performed on samples.

In the case of industrial activities involving radioactive substance manipulation, area radiation monitors are extensively used to ascertain the integrity of equipment containing radioactive products.

F. Status of the Process Fluid

A concern for degradation originates also from the chemical or radiological status of the process fluid. The chemical status can strongly influence the advance of corrosion in equipment material or can be indicative of a loss of integrity in upstream containment barriers. The second case pertains typically to the nuclear power plants; an increase of radioactivity in the reactor coolant can derive from a loss of integrity in nuclear fuel, that can be confirmed by a coolant composition analysis indicating the presence of fission products in the coolant.

The analysis of process fluid is made by means of sampling stations to which a suitable amount of fluid diverted by the process is continuously conveyed and monitored.

Conductivity and pH are common parameters to be monitored. Turbidity is also worthy of detection as it correlates directly with the rate of generation of corrosion products. Instrumentation channels are involved for detection, indication, recording, and alarm.

For a deeper knowledge about the process fluid chemical composition, laboratory measures can be executed based on chromatography or spectrography techniques.

G. Remotized Visualization

Direct observation of equipment is worthy of integrity assessment. Due to accessibility problems (environment conditions or particular location of equipment), serious limitations can be posed to direct inspection; closed circuit television systems are then used for remotized visualization. A number of movable transmitting cameras is located in the equipment area; the receiving screen and the system controls are placed in a most favorable location. The system control includes the selection, the positioning, and the orientation of the movable cameras. The cameras can be shaped so to enter into equipment for interior visualization.

IV. DIAGNOSTIC INSTRUMENTATION

The purpose of diagnostic instrumentation is to collect and record significant plant events and variables, to present to the operator detailed and complete information about the status of processes and components, and to support the operator in the analysis of performance in the medium-long term. Typical information provided by diagnostic systems to the operator are referred to safety margins and to deviations with respect to the best theoretical operating conditions so that the operator can be quickly alerted about abnormalities and can be predisposed to take adequate provisions to bring back the plant operation to the desired quality and safety level.

Furthermore, events recording (referred to operational and accident transients) provides useful means for the improvement of operating conditions, control logics, component selection, and preventive maintenance programs.

The architecture of diagnostic instrumentation is based typically on the utilization of a process computer supported by a number of peripheral units for data recording (magnetic tapes or disks, punched tapes, etc.) and presentation to the operator (screens, graphic units, printers, mimic panels, etc.). A set of software programs is devoted to the plant data acquisition and elaboration, periodically or on operator demand. Main aspects characterizing this architecture as a diagnostic system for a specific plant are component performance, system flexibility, and human factor engineering.

Recording units allow a large amount of plant data to be memorized, conserved, transported, and recovered.

Presentation devices constitute, with regard to the operator, flexible interactive interfaces characterized by a high capability of information representation.

Programmability allows application programs to be extensively used and modifications to be easily implemented.

Flexibility allows to adopt system potentialities to the specific plant diagnostic requirements. Human factor engineering takes into account and optimizes the interface between the operator and the presenting devices. Some examples of utilization of diagnostic systems are outlined below.

A. Verification of Plant Operational Parameters
Two main aspects can be pointed out in this area:

1. Plant operation survey through the entire set of the monitored variables
2. Evaluation of specific operation aspects through calculated parameters

With reference to plant survey a major concern is for organization of information transfer to the operator.

The effectiveness of the diagnostic system depends on its capability of providing information with the due respect of exigencies of synthesis and correlation about data to be presented as a whole. A practical case consists in the use of CRT (cathode ray tubes) as process computer terminals showing schematically plant systems and related variables in the form of mimics. A mimic is a representation on the screen of major process lines and components with the indication of both the current variables values and reference values (operational and safety limits); provisions are implemented to put in evidence abnormal status (flashing, color coding). Mimics can be organized hierarchically to make available informations from plant level to system and component level with an increase of details. Another kind of representation on the screen is the evolution in the time of selected plant variables, which allows to be aware of the global status and trend.

A different area for computer-based diagnostic systems is the on-line calculation of plant parameters which cannot be monitored directly by monitoring instrumentation; a practical example is the calculation of fuel thermal performance parameters in nuclear power plants. These parameters are important to evaluate the mechanical integrity of the nuclear fuel and the potential for fission product release to the coolant. The operator can be sure about fuel integrity if certain thermal parameters stay into predetermined acceptable values. As the parameters cannot be physically measured with sensors, a derivation by other measurable process variables must be accomplished by means of mathematical models and algorithms which are organized on the basis of the physical laws governing the thermal phenomena. The plant process computer is the natural means to be employed by the operator for calculation and presentation of these parameters.

B. Safety Parameter Display System

The safety parameter display system (SPDS) is a diagnostic system used in nuclear power plants to help operating personnel to assess quickly safety status, providing concise information on critical plant parameters.

The displayed parameters are associated with safety functions (such as reactivity control, emergency cooling, integrity of barriers, and radioactivity control).

SPDS is used in addition to other controls available in the control room and allows the operator to make safety assessment in a timely manner without surveying the entire control room. SPDS provides continuous indication and recording during normal and abnormal conditions; this performance can be particularly important during transients and the initial phase of an accident. Upon the detection of an abnormality in the plant, additional information is provided for a deeper diagnosis (including the consequences of the abnormality) and for assistance in the selection of mitigative actions.

Design of SPDS includes provisions to prevent faulty processing or failed sensors from compromising the quality of the provided data; when practical all data are validated on a real-time basis before being displayed. For example, data from redundant sensors are compared for congruence, the values of parameters checked against predetermined limits.

A typical computer-based SPDS includes data acquisition, control and storage, and presentation.

Data acquisition takes the data from plant sensors by means of isolation devices if sensors are concerned with safety functions and their integrity must be assured even against SPDS faults. Data are sorted and duplicated (if they have also other uses, such as alarm) and are sent to the control and storage subsystem through multiplexer, formatter, and A/D converter.

Data storage and elaboration needed for SPDS functions are accomplished by means of one or more computers. The data presentation subsystem provides all the man-machine interface functions of the SPDS. It is usually constituted of CRT displays, printers, keyboards, graphic terminals, and any other unit needed to give information or to receive commands to/from the operator and the engineer staff of the plant.

C. Expert Systems

The high development in the electronic technology of recent years has made possible a new kind of diagnostic system based on the application of the artificial intelligence, and an increasing diffusion of such a system can be foreseen for the near future. The main goal pursued with this system is to provide the operator with a powerful means which, in addition to the typical functions performed by a conventional diagnostic system, is capable to lead him in the conduction of the plant and in the analysis of its performance.

This characteristic can be essential during accident conditions in which the need for precocious diagnostic and safe and sudden actions by the operator is increased, while at the same time the operator's effectiveness decreases because of the necessity of correlating promptly the great amount of unusual data generated by the plant.

A practical example can be the automation of emergency procedures; these procedures are established by design on the basis of abnormalities and accidents which are anticipated to affect the plant, and should be implemented by the operator following and in correlation with preestablished symptoms and/or events.

This correlation can be managed by an expert system in selecting automatically the most adequate procedure for any given circumstance; assistance can also be provided for the correct implementation of the procedure by the operator.

Another practical case is diagnostic for complex systems or components; an expert system can store all available technical knowledge about phenomena concerned with operation/maloperation and can provide rapid and reliable information about the status.

A prospect for the future is to use an expert system as a support to increase the technical knowledge rather than to optimize its employment. By a technological point of view, the diagnostic expert systems are characterized by a powerful process computer with a large storing and computing capability and, over all, by a software which reproduces the knowledge and the way of thinking of a highly skilled operator, and by screen-printer output units with high interactive performance with regard to the operator.

V. ENVIRONMENTAL MONITORING FOR HIGH-RISK PLANTS

Major danger and damage concerned in high-risk plants are connected to the release of dangerous substances from equipment and areas of the plant where they are confined normally to the environment.

Two main factors addressing the possibility of release can be pointed out. First, the means adopted for the confinement can fail because of degradations and accidents, and significant releases (accident conditions) to the environment will result. Second, design provision and technological means employed for the confinement of dangerous materials can be limited in itself, so that a perfect confinement is not attainable and some release which cannot be reduced below certain values will result (operational releases).

Both operational and accidental releases raise the problem of evaluating the possible consequences for the operating personnel, the public, and the environment (besides the economic consequences). Another concern involved is the arrangement of mitigative actions in the plant to contain the entity of releases and of consequences.

Finally, consideration must be taken realistically for accidents involving large and uncontrolled releases, in which drastic emergency actions, such as evacuation of operating personnel and neighbors, could be required.

In all these cases it is essential to have a measure of the entity of the releases and the contamination. The reasons which dictate specifically this exigence include the following:

1. Need of checking continuously if the operational releases stay below acceptable values to confirm the persistence of normality
2. Need of verifying (tempestively) the necessity of mitigative actions on the plant (to be performed automatically or based on operator judgment)
3. Need of evaluating the amount of release and environmental contamination to assess the consequences
4. Need of evaluating tempestively the opportunity of actuating an emergency plan for the health protection of population

An environmental monitoring system strongly helps to reduce the consequences of the contamination. The extension of this system concerns:

1. The internal area of the plant which can be the first area to be interested by the releases

2.   Predetermined points of release to the exterior of the plant
3.   External environment
4.   Meteorological conditions, which can influence the diffusion of the released material into the environment

The parameters to be measured by the monitoring instrumentation are referred to the recognition of the contamination in itself (or its effect, such as radiation level in the case of radioactive contamination) and to other physical factors which can influence the entity of the release or its diffusion in the environment. These parameters can be exemplified as follows:

1.   Concentration of contaminants in the environment inside and outside the plant (discharging points, atmosphere, water, ground, exposed surfaces, and food)
2.   Radiation level caused by the exposure of radioactive sources
3.   Flow and contamination of effluents in discharging points and duration of discharge
4.   Wind direction and speed and vertical temperature difference in the atmosphere (estimation of atmosphere stability)

A. Monitoring Methods and Instrumentation

1. Monitoring Inside the Plant

The presence of contamination inside the plant can be revealed by means of detectors distributed in the areas to be monitored. The detection is based on chemical or physical properties of the contaminants, e.g., the ionizing effects, the light absorbtion characteristic, and the catalytic combustion effects.

The local detectors work continuously supplying signals to a centralized monitoring system; they can also be equipped with individual alarming features to warn the personnel who stay in the influence area of each device.

Another way of revealing contaminants is by sampling and measuring stations; samples are collected in turn from different points and are sent to a chromatographic analyzer; this arrangement can be conceived for a fully automated operation. Samples can also be taken for laboratory analysis which are based on spectrographic or chromatographic methods.

2. Effluents Monitoring

Effluents consist in the discharge of wastes from the plant to the external environment through chimneys, vents, and draining lines. Contaminants contained in the waste should be carefully monitored due to their impact on the external environment. The amount of contaminants released can be controlled by means of flow and concentration measures in the point of release.

Concentration can be monitored with detectors on a continuous basis or periodically by sampling and analysis. Flow of effluents can be measured with rotameters or by means of calibrated orifices.

3. Meteorological Monitoring

Wind characteristics (speed and direction) are measured with anemometers. Atmosphere temperature is measured in different points along a vertical line by means of temperature sensors to assess the atmospheric stability as it affects the release dispersion.

4. Monitoring Outside the Plant

In the external environment the monitoring of contaminants is generally performed by means of laboratory analyses. The samples to be analyzed are collected from selected locations and consist of amounts of ground, water, air, food, and vegetables. These samples can be analyzed by means of chromatographic or spectrographic methods.

B. Environmental Monitoring for Nuclear Power Plants

The parameters to be monitored include the concentration of radioactive substances, the isotopic composition, and the radiation type and level. The radioactive contamination can interest the atmosphere in the form of particulates, halogens, and noble gases.

A method to measure the particulates is by filtering a sample of atmosphere and measuring the radiation level of the filter. This level corresponds to the amount of particulate in the filter which depends on its concentration in the original sample.

The isotopic composition can be measured with the γ-spectrographic method which is based on the isotopic differences of the radiation energy. The radiation is measured with Geiger-Muller detectors, scintillation detectors, and ionization chambers.

A typical radiation monitoring system includes several detectors located in points of the plant with significant radiation levels. The detectors have the capability of generating locally sound and visual warning signals and alarms; furthermore, they send signals for indication and recording in the control room, where alarms are also repeated. The detectors are also equipped with self-diagnostic circuits to reveal malfunctions.

The variables to be monitored to determine the entity of the release of radioactive materials are listed as follows:

1. Radiation inside the plant buildings
2. Concentration of radioactive material in effluents released from the plant to the external environment; it includes noble gases, particulates, and halogens
3. Concentration and radiation level in the environs
4. Meteorological status (wind speed and direction and atmosphere stability)

The characterization of the monitoring instrumentation covers the following aspects, according to the exigencies posed by the plant for normal and accident conditions: redundancy, measuring range and accuracy, alarm and trip setpoints, independence and diversification, calibration and maintenance, location of detector readout alarm sampling points and sampling stations, provisions for purging sampling lines, correlation between readouts sampling results, and plant operation procedures.

The environmental monitoring for nuclear power plants includes, as a particular case, the seismic monitoring; this is not correlated with the release and radiation monitoring but it is important to detect a seismic event and to initiate safety actions following an earthquake. The seismic detectors consist generally of triaxial inertial accelerometers placed on the ground in the proximity of the plant.

C. Regional Data Network

An environmental monitoring system cannot be thought to operate singularly and simply as an expansion of the plant monitoring systems covering the plant proximity. It needs to be integrated into a larger scale survey system, conceived on a regional basis for a comprehensive management of emergency situations in front of factors such as the following:

1. Incidence of the distance from the plant on the effects of releases
2. Effects of metereological conditions on the diffusion of the releases into the atmosphere (direction of propagation imposed by winds) and on the deposition of dangerous material on the ground
3. Necessity of being aware of the integrated exposition degree of population resident farther than in the plant proximity
4. Necessity of reassuring neighboring population not actually exposed

All information referred to the conditions of the failed plant, the environment, and the health situation should be transferred to a regional data network based on a dedicated communication system. This network should report all information to a central safety administrative office equipped with sufficient automated computing means to permit rapid decisions be taken by the responsible authorities.

The network should also be conceived to cover all the emergencies (nuclear, chemical) which could arise from the high-risk plant located in the region.

A permanent connection should be maintained, by means of the network, between plant and each monitoring station in the environs and the central safety office. A permanent connection to the intervention people (such as firemen) should also be maintained to assure the actuation of emergency action with the due rapidity.

REFERENCES

1. Standard Criteria for Protection Systems for Nuclear Power Generating Station, ANSI/IEEE 279-1971, Institute of Electrical and Electronics Engineers, New York, 1971.
2. Standard Criteria for Independence of Class 1E Equipment and Circuits, ANSI/IEEE 384-1981, Institute of Electrical and Electronics Engineers, New York, 1981.
3. Standard for Qualifying Class 1E Equipment for Nuclear Power Generating Stations, ANSI/IEEE 323-1983, Institute of Electrical and Electronics Engineers, New York, 1983.
4. Standard Criteria for Type Tests of Class 1E Modules Used in Nuclear Power Generating Stations, ANSI/IEEE 381-1977, Institute of Electrical and Electronics Engineers, New York, 1977.
5. Standard for the Design and Qualification of Class 1E Control Boards, Panels and Racks Used in Nuclear Power Generating Stations, ANSI/IEEE 420-1982, Institute of Electrical and Electronics Engineers, New York, 1982.
6. Degrees of Protection of Enclosures for Low Voltage Switchgear and Controlgear, IEC 144, International Electrotechnical Commission, Geneva, 1963.
7. Software for Computer in the Protection System of Nuclear Power Generating Stations, IEC 880, International Electrotechnical Commission, Geneva, 1986.
8. Instrumentation for Light Water Cooled Nuclear Power Plants to Assess Plant and Environs Conditions During and Following an Accident, RG 1.97, U.S. Nuclear Regulatory Commission, Washington, D.C., 1983.
9. Electrical Instruments in Hazardous Atmosphere, ISA RP12.1, Instrument Society of America, Research Triangle Park, NC, 1960.
10. Functional Criteria for Emergency Response Facilities, NUREG 0696, U.S. Nuclear Regulatory Commission, Washington, D.C., 1981.

Section IV
Exposure and Consequences Limitation

INTRODUCTION

In Section IV and Section I of Volume II the main safety criteria to be considered by the designer of a plant handling hazardous substances have been analyzed.

Such criteria are aimed at avoiding damages to the plant, thus avoiding releases of the hazardous substances.

The safety philosophy cannot be limited, nevertheless, to the prevention aspect; the relevance of consequences of releases of high-risk substances imposes the consideration of accidents that — in spite of the preventive measures adopted in the design — might however occur.

In addition, several processes where noxious substances are handled are characterized by releases not limited to accidental conditions but during normal operation (continuous or batch). It is, therefore, important to be able to evaluate the modes of dispersion of effluents into the environment, so as to be able to evaluate the exposure for humans, plants, and animals. The knowledge of the physics of dispersion of harmful substances allows the identification of the consequences of the normal or accidental releases; it is the basic point for the study of actions able to limit or reduce the consequences of releases that in any case cannot be reduced, so as for the identification of maximum values of releases acceptable during normal operation or to hypothesize as consequences of accidents (and to which to bind the design itself), if limits on the doses (more generally, on the consequences) are fixed.

Chapter 10

EFFLUENTS DISPERSION AND METHODS TO EVALUATE THEIR EFFECTS

L. Bramati

TABLE OF CONTENTS

I. INTRODUCTION

Of the risks associated to industrial plants many are related to the concept of environmental dispersion (both inside and outside the plant) of the dangerous substances treated by the same plant either as part of the end products or as processing byproducts. It is essential to stress this point in order to identify the dispersion risk for toxic or dangerous substances used in the various types of plant; a primary chemical industry processing toxic elements such as chlorine or phosgene is automatically classified as a risk plant, hence, it is compulsory to carefully evaluate the effects on man and the environment of the plant dispersion. On the other hand, in the case of secondary chemical plants commonly using harmless substances or plants using small quantities of dangerous materials for secondary purposes, the chronic or accidental dispersion risk is often neglected. For example, think of the small industry in Seveso and its dioxin contamination or the elimination of old thermal plants which had been insulated with a considerable amount of asbestos.

For this reason, it is extremely important to make a detailed list of all the raw materials, processed products, equipment, structural components, and manufacturing byproducts used in the plant.

Each of these elements shall be classified according to quantity and physical shape in order to evaluate their possible airborne or liquid environmental dispersion. Such evaluation should indicate whether an accidental or chronic dispersion (due to the normal activity of the plant) would be possible or not.

This analysis is the first step to describe the possible scenarios of diffusion and its environmental impact and is the starting point for the application of methodologies listed below. The evaluation of consequences should be as much detailed as possible, according to the predictable risk; of course, it is necessary to follow an iterative process made up of sequential phases each one being more precise and realistic than the previous ones. The imperfect knowledge of diffusion processes and the various environmental chains implies the use of many safety factors during the preliminary evaluations in order to get a truly reliable prediction of risk. On the other hand, a good technical and experimental knowledge of environmental diffusion processes, often achieved with considerable efforts in terms of economic and human resources, will allow to employ sophisticated models and more realistic scenarios with the consequent result of reassuring risk evaluations. The process will stop when either consequences are considered intrinsically unimportant or prevention and intervention means are considered sufficient both in terms of commitment and efficacy.

II. DISPERSION OF AIRBORNE EFFLUENTS

A. General

The dispersion of effluents present in the atmosphere depends upon several parameters such as wind direction and speed, atmospherical turbulence, air and effluent temperature, ground characteristics, type of plant, emission modalities, and chemicophysical characteristics of effluents.

It is obviously a very complex process that should be simplified in order to have a realistic mathematical simulation. The literature is full of references of this kind, but an overall description of the subject is out of our present scope. References include some books which systematically discuss the subject.[1-3]

Thus, we will limit ourselves to a synthetic approach which aims at assessing the importance of individual diffusive phenomena; it should be noted, however, that every theoretical evaluation of effluents concentration, far from the source, is affected by a certain measurable degree of uncertainty.

Table 1
STABILITY CLASSES ACCORDING TO WIND SPEED (M/S) AND LAPSE
RATE (°C/100 M)

Wind speed U	ΔT/ΔZ						
	< −1.5	−1.1 + 1.2	−1.1 + 0.9	−0.8 + 0.7	−0.6 + 0	0.1 + 2	>2
<1	A	A	B	C	D	F	F
1 < V < 2	A	B	B	C	D	F	F
2 < V < 3	A	B	C	D	D	E	F
3 < V < 5	B	B	C	D	D	D	E
5 < V < 7	C	C	D	D	D	D	E
7 < V	D	D	D	D	D	D	D

B. The Main Meteorological Parameters which Influence Diffusion

The main meteorological parameters which influence diffusion are the wind direction and speed and the atmospheric turbulence.

At the emission height of effluents, wind speed has an inverse and linear effect on their concentration. Through direct measurements, we often know speed (U_o) at height (Z_o), which differs from speed (u) at emission height z. The equation which links the two velocities is as follows:[2]

$$u = U_o \left(\frac{z}{Z_o} \right)^P$$

where p is a parameter ranging between 0.12 and 0.40, according to the atmospheric turbulence.

Wind direction is taken as constant, but in case of actual releases with possible health consequences, it would be advisable to constantly verify the real direction of wind both at ground and at the emission height; the two air layers could take part in different wind currents, even showing divergent patterns, because of the well-known shear phenomenon.

The atmospheric turbulence is usually treated according to the Pasquill-Gifford method which considers six different turbulence categories, from A to F; it comprises the whole turbulence spectrum, from unstable to stable, according to the following pattern:

- A: very unstable
- B: moderately unstable
- C: slightly unstable
- D: neutral
- E: moderately stable
- F: very stable

There are various methods to classify and evaluate the degree of stability, based on other types of meteorological observation. The two most common methods take into account the wind waving σ_θ and the lapse rate.

The first method is based on the obvious observation that the atmospheric turbulence directly affects the wind velocity fluctuation, measured by an anemometer. We can measure the SD σ_θ or side waving of wind with a special wind vane in order to define the stability category from A to F, where σ_θ varies from 25° to 2.5°, in steps of 5° from one category to the other (apart from E to F, where the step is of 2.5° only).

As far as the second method is concerned, atmospheric turbulence depends upon kinematic (wind speed) and thermal forces (lapse rate). For this reason, Table 1 was proposed; it defines the stability category as function of the average wind speed and the temperature difference over a 100-m height.

It should be remembered that these methods are semiempirical. On the other hand, an instability error of one step does not entail marked errors in the diffusion assessment which is known with a factor 10 of uncertainty.

C. Source Factors which Influence Diffusion

Apart from meteorological factors, diffusion is mainly influenced by the chimney size (width and height), emission speed of effluents, and their temperature. According to ASME methodology,[2] two cases are possible: almost cold (<50°C) and hot emissions (>50°C).

In the first case, the effluent rise over the chimney top is basically determined through the discharge speed and can be calculated as follows:

$$H = D\left(\frac{V_s}{u_s}\right)^{1/4}$$

where H (m) is the source rise from the chimney stack top, D (m) is the internal diameter of the stack, Vs (m/s) is the efflux speed, and u_s (m/s) is the wind speed at the chimney top.

As for hot effluents, the discharge rise is mainly affected by the buoyancy due to the different density of effluents compared with ambient air. The phenomenon varies according to the lapse rate, i.e., atmospheric stability.

For stable conditions (D,E,F) we can apply the formula

$$H = 2.6 \cdot \left(\frac{F_s}{u_s\, S}\right)^{1/3}$$

where F_s = buoyancy push = $g \cdot V_s \left(\frac{D}{2}\right)^2 \cdot \frac{\rho_e - \rho_s}{s} \left[\frac{m^4}{S^2}\right]$, g = gravitational acceleration (m/s²), ρ_s and ρ_e = density of gases and environmental air (g/m³), S = stability parameter $= \frac{g}{T}\frac{\partial\theta}{\partial z}$, T = air ambient temperature (K), $\frac{\partial\theta}{\partial z}$ = temperature potential gradient $= \frac{\Delta T}{\Delta z} - \Gamma$, $\frac{\Delta T}{\Delta z}$ = ambient temperature lapse rate, and Γ = dry-adiabatic lapse rate = 0.0098 $\frac{K}{m}$

In neutral or unstable conditions (A,B,C) we can apply the formula

$$H = \frac{1.6\, F^{1/3} \cdot (3.5\, X')^{2/3}}{us}$$

where $X' = 14\, F^5 \nabla^8$ if F < 55 and $X' = 34\, F^2 \nabla^5$ if F > 55.

Another factor to be considered is the release height compared with that of the surrounding buildings. When the ratio between the two values is less than 2.5, the turbulence of buildings causes a discharge carryover within the vortex produced by the same buildings; hence it can be assumed that the discharge takes place at ground level, having an initial transverse size of the plume comparable with the building cross-section.

As for the large number of simplifications adopted in the analysis of diffusion, we should keep in mind that the ground is assumed as smooth and flat; buildings, hills, and valleys may often remarkably change to the diffusion mode, although we do not have sufficiently sophisticated models capable of simulating this complex reality; in this latter case, theoretical calculations shall be used with extreme caution, taking into account proper safety margins to be inferred from empirical observations or previous experiences.

D. Diffusion Coefficient

Having defined atmospheric conditions and emission modalities, we can now calculate the concentration at ground level of an effluent in a given point. The diffusion coefficient is defined by the following expression:

$$\frac{X}{Q}(x,y) = \frac{1}{\pi\sigma_y\sigma_z u} \exp - \left[\frac{H^2}{2\sigma_z^2} + \frac{y^2}{2\sigma_y^2} \right] \tag{1}$$

where X/Q = diffusion coefficient (s/m³), σ_y and σ_z are the lateral and vertical SD of the plume (m), u (m/s) is the wind speed at the stack outlet, H is the actual height of release (height of the chimney + elevation), and y and x are the coordinates of the point of interest as compared to the origin (point of release).

If we define Q as the discharge rate (e.g., kg/s), $X/Q \times Q$ will be the actual concentration expressed in kg/m³; if we know the total discharge (e.g., kg), $X/Q \times Q$ will give the kg s/m³, time integrated concentration. Despite its unclear physical significance, this parameter is extremely useful to calculate the exposure of people to pollutants in case of accident. For example, given the breathing rate R (m³/s), X·R will represent the inhaled quantity throughout the accident.

In fact,

$$\frac{Kg\ s}{m^3} \times \frac{m^3}{S} = \text{total number of kg inhaled}$$

(In practice, they are always very small fractions of kilograms!)

Expressions concerning particular cases can be deducted by Equation 1:

If we want to know the concentration at ground along the plume axis we obtain:

$$\frac{X}{Q} = \frac{1}{\pi\sigma_y\sigma_z u} \exp - \left[\frac{H^2}{2\sigma_z^2} \right] \tag{2}$$

When emission is at groundlevel, concentration is given by:

$$\frac{X}{Q} = \frac{1}{\pi\sigma_y\sigma_z u} \exp - \frac{Y^2}{2y^2} \tag{3}$$

while along the plume axis we obtain

$$\frac{X}{Q} = \frac{1}{\pi\sigma_y\sigma_z u}$$

The maximum concentration value is approximately given by:

$$\left(\frac{X}{Q}\right)max = \frac{2}{\pi e u H^2} \cdot \frac{z}{\sigma_y} \tag{4}$$

where e = 2.718.

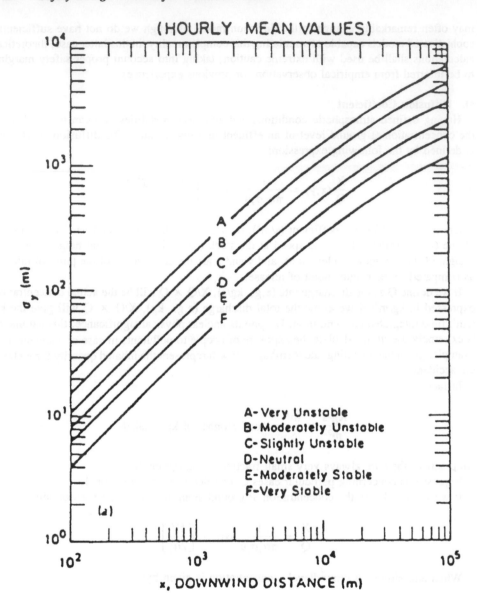

(HOURLY MEAN VALUES)

A-Very Unstable
B-Moderately Unstable
C-Slightly Unstable
D-Neutral
E-Moderately Stable
F-Very Stable

σ_y (m)

x, DOWNWIND DISTANCE (m)

FIGURE 1.   Horizontal SDs of a plume. In the dispersion equations, the crosswind plume SDs are functions of distance (x) and meteorological conditions. Six such plots are shown. The curves are tentative beyond 10,000 m, since only limited experimental data have been collected beyond that distance.

From this formula it results that, for the emission at groundlevel (H → 0), the maximum concentration corresponds to the singular point which is represented by the point of emission.

As it can be noticed, parameters requiring a conceptual and numerical definition are SDs of the plume, σ_y and σ_x. Formulas and semiempirical methods have been developed to this purpose; in fact, it is clear that for the same type of stability we can have a quite wide range of values referred to the plume SD, according to the type of ground and the local atmospheric turbulence. We can simplify the problem even in this case through some σ_y and σ_z standard diagrams, related to distance and stability class. Figures 1 and 2 show ASME diagrams.[2] Reference 3 illustrates a series of values depending upon the ground "roughness".

The above formulation applies to diffusion extending over a limited period of time but

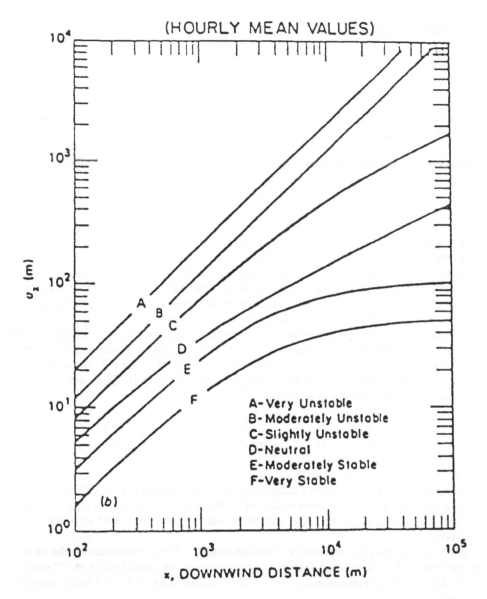

FIGURE 2. Vertical SDs of a plume. The stable plots approach constant values beyond 10,000 m. It is believed that vertical dispersion almost ceases in such conditions, provided there are no breakdowns of the stable layer by strong wind shears. The curves are tentative beyond 10,000 m.

can also be easily applied to the calculation of the average concentration over long periods, e.g., for the yearly average; in this case, it will be necessary to know the wind persistence percentage for each sector, wind speed, and stability class, i.e., the wind rose. Then, a further simplification which mediates the wind on the side parameter (y) within every sector of the wind rose (usually 22.5, 30, and 45°) is sufficient to our purpose. For example, in the case of 22.5° sectors, Equation 1 becomes:

$$\frac{X}{Q} = 2.032 \cdot \sum_{ij} f_{ij}[Xu_{si}\sigma zj]^{-1} \exp\left[-\frac{H^2}{2\sigma_y^2}\right] \tag{5}$$

where (X/Q) = mean concentration factor $(\frac{s}{m^3})$,

$$2.032 = \left(\frac{2}{\pi}\right)^{1/2} \cdot \left[\frac{2\pi}{16}\right]^{-1} \tag{6}$$

f_{ij} = wind frequency for i velocity class and j stability class, x = distance used for calculating concentration (m), u_{si} = i class of wind speed (m/s), σzj = standard vertical deviation for distance x and j stability class (m), and H = release height (it might also be H = 0) (m)

According to this formula, each class of stability should include different wind frequencies according to the various speed ranges: but it may be that meteorological data available are less sophisticated and in such a case we just have to eliminate i index.

E. Deposition

In many cases of atmospheric dispersion of radioactive or toxic contaminants in the form of aerosol, these substances are deposited on various surfaces: on the ground, on vegetables, animals, and even people. This long-lasting phenomenon should be taken into account especially in the case of short accidental episodes. The analytical discussion of the problem is very complex, since many factors are involved: the diameter and density of particles, the physicochemical interaction with surfaces, etc.

In order to simplify the problem it is necessary to analyze only the diffusion of true aerosols, i.e., particles with a diameter less than 10 μm; larger particles are assumed to have a fast sedimentation in the first hundred meters around the source (ballistic trajectory). Aerosols are thought to have a theoretical deposition velocity (vg). Thus, the surface contamination can be expressed as follows:

$$\frac{W}{Q} = Vg\frac{X}{Q} = Vg \cdot \frac{1}{\pi\sigma_y\sigma_z u_s} \exp - \left[\frac{Y^2}{2\sigma_y^2} + \frac{H^2}{2\sigma_z^2}\right] \tag{7}$$

where W (x,y) has the size of m^{-2}.

Both Equations 1 and 7, when multiplied by Q total discharge (kg, or Ci) give the actual surface contamination, while when multipled by the emission rate (kg/s or Ci/s) provide the surface contamination for the time unit to be subsequently integrated over the time of emission.

We have described the so called dry "contamination". If we want to identify the localized phenomenon of contamination following precipitations, we should adopt the "washout" coefficient which is represented by ∧ parameter instead of vg. It is used in the formula as follows:

$$\frac{W}{Q} = \frac{\wedge}{uy(2\pi)^{1/2}} \exp - \frac{Y^2}{2\sigma_y^2} \tag{8}$$

where ∧ has the size of T^{-1}.

This formula applies to a rainfall limited at the x point. When the precipitation affects all the area included between the source and x, we shall consider the cloud depletion through the expression

$$\frac{W'}{Q} = \frac{\wedge}{\mu_s \sigma y \cdot (2\pi)^{1/2}} \cdot \exp - \frac{Y^2}{2\sigma_y^2} \cdot \exp - \frac{\wedge x}{u} \tag{9}$$

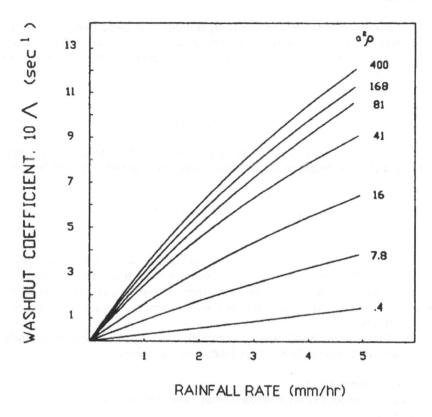

FIGURE 3. Washout coefficient for unit density particles vs. rainfall rate and $a^2\rho$.

$\wedge$ Parameter depends upon aerosol size and its chemico-physical characteristics as well as upon the intensity of the rainfall.

It varies from 10^{-3} to 10^{-5} (s^{-1}) for precipitations between 10 and 0.1 mm/h. To this end, Figure 3 is taken as an example.

F. Special Meteorological Situations

Even a rough discussion about special meteorological situations would go beyond the scope of this short presentation. Thus, we will only mention two particular cases for which classic methods are not applicable (see literature for further details).

Fumigation or intrapment of the plume — This condition occurs when an atmospheric inversion layer avoids vertical diffusion beyond L height level. In this case we use the formula:

$$\frac{X}{Q} = \frac{1}{(2\pi)^{1/2}\sigma_y uL}$$

Fog and calm — For this case, when the air movement is not recorded by common anemometers and there is no indication as to the movement direction, some authors[4] have hypotized a homogeneous diffusion in all directions at a fictitious speed which should indicate the intrinsic diffusion rather than a real dynamic speed.

The case of calm, mainly in F stability category, is treated with the following formula:

$$\frac{X}{Q} = \frac{1}{2\pi r^{3/2} \cdot (u_d \cdot K_2)^{1/2}} \qquad (10)$$

where r is the distance from the source (m), u_d is a diffusion speed (which can vary from 0.3 to 0.5 m/s), and k_2 is the coefficient of vertical diffusion (having a value of 1 m²/s).

On the other hand, we have the mist case that entails a net layer of H (m) height inversion. It can be modeled through the formula:

$$\frac{X}{Q} = \frac{1}{2(2\pi)^{1/2}\alpha rH} \tag{11}$$

where α is a constant similar to diffusion speed (0.5 m/s), r is the distance from the emission point (m), and H is the height of the inversion layer and is usually included between 100 and 300 meters (m).

III. DIFFUSION OF LIQUID POLLUTANTS

A. General

As far as the risk evaluation is concerned, water diffusion of radioactive and toxic pollutants is easier to study than the atmosphere. This does not mean that the hydrosphere can be modeled more easily than the atmosphere; more simply, factors such as speed, diluent masses, and better confinement, make water-related risks less severe than the risks connected to atmospheric releases.

From the diffusion standpoint, we will separately discuss three cases: rivers (including small lakes as well), seas (including large lakes), and the subsoil.

B. Diffusion Through Rivers

A discharge, either continuous or discontinuous, into rivers is diluted in two quite separate sequential phases

1.   In the initial plume, until all the river flow rate (near field) is involved
2.   In the following sections, through dilution of influent streams only (far field)

As far as the near field is concerned, there are quite complex models to evaluate the plume shape although they rarely achieve satisfactory results, because of the problem of analytically simulating obstacles, bottoms and shores, etc. When necessary, physical models in scale can be used.

In the case of the risk evaluation, we have to quantify the order of magnitude of the distance by which Co initial concentration is transformed into C, i.e., the river mean concentration. Also in the case of effluents with a Co concentration only slightly different from C, such a distance can be as long as many scores of kilometers because of the absence of obstacles, bends and weirs. Discharge shape, efflux velocity and water turbulence are extremely influent but difficult to simulate.

If Fd = Co/C is the coefficient of dilution, it is also proportional to P/Po, where P and Po are the river and discharge flow rates (always <1 and in almost all cases ≪ 1).

When required by the environmental analysis, the problem can be studied assuming the extreme case Fd = 1, i.e., in the immediate vicinity of the discharge point.

In a situation of equilibrium, when the whole river is involved in the dilution process, the concentration of a continuous discharge is given by:

$$C = \frac{C_0}{F_d} = C_0 \frac{P}{P_0}$$

However, it should be clear that the river flow rate is often extremely variable and data on low rates are more subject to errors although it is this kind of situation which is extremely dangerous from a contamination point of view. In fact, water gauges to measure the flow rate are normally designed to measure large rates. Moreover, the flow rate is basically given by the height of the water head as compared to the bottom of the river; when the flow rate decreases, the height tends to zero thus making the measurement uncertain because of the nonuniformity of riversides and riverbed. According to the type of release, we take into account instantaneous or mediated rates over more or less long periods of time. However, we should bear in mind that every part of a river has its own characteristic, ranging from a brook status in the vicinity of sources, to channel conditions in the final sections or sections including artificial barriages.

In order to calculate the river concentration, we should also take into account suspended matter in the water course and its physicochemical interaction with pollutants discharged.

In particular, the clayish material suspended in water often presents a strong affinity with some elements, typical of inorganic ion exchangers.

The following sedimentation of this material, when the current slows down or when the river suddenly deviates, creates some artificial areas of high concentration which is displaced by the following floods and should be carefully considered in the risk analysis.

C. Diffusion Through Seas

Similarly to rivers, the sea discharge diffusion takes place in two phases: first a turbulent dilution until the kinetic energy of the discharge is exhausted and then a slow dilution due to the current, wave motion, and diffusion.

Even in this case, some complex three-dimensional models can simulate the effects of the bottom, the motion field of currents, etc. When no analytical data on currents and wind effects are available, it is advisable to use simple models that can be run on PC to make the risk contamination assessment.

The simplest case is when the effluent has a quite large flow rate, with a possible risk exposure which could occur before any remarkable dilution into the sea. In this case, an empirical dilution factor for short distances can be applied to Co initial concentration.

When the discharge has lost its initial thrust, it is possible to simulate diffusion by means of a quite simple model proposed by Brooks;[5] it was used to study the dispersion of Los Angeles urban waste into the ocean starting from a discharge site located at some miles from the coast.

The basic point of this method is that ϵ diffusion coefficient depends upon the transverse size of the flux, L, according to the law

$$\epsilon = \alpha L^{4/3}$$

where α is a constant equal to 0.01 when units are expressed with CGS system. ϵ Behavior is close to experimental data, as it can be seen from Figure 4. The concentration in an x,y point can be calculated through the expression:

$$C(x,y) = \exp{-\frac{\lambda x}{V_x}} \cdot \frac{C_0}{2} \cdot \left[\operatorname{erf} \frac{y + \frac{b}{2}}{\sigma} - \operatorname{erf} \frac{y - \frac{b}{2}}{\sigma} \right]$$

where Co is the initial concentration, b is the initial width of the discharge, Vx is the current speed, λ is a disappearance speed (decay, chemical reaction, etc.), erf is the error standard function:

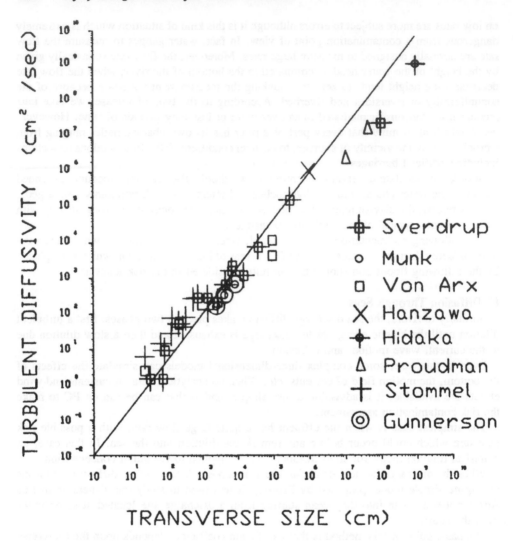

FIGURE 4.   Side diffusivity: experimental data and Brooks' law.

$$\frac{2}{(\pi)^{1/2}} \int_0^x e^{-V^2} dV$$

x is the coordinate along the flow axis, and y is the lateral coordinate:

$$\sigma = \left[\frac{b^2}{\sigma} \left(1 + 8\, \epsilon_0\, \frac{x}{b^2} Vx \right)^3 - 1 \right]^{1/2}$$

Figure 5 gives a better description of the way used to discuss the problem.

A different approach to diffusion through the sea is possible when we have well-limited bays or lagoons; given the size of the basin and assuming a slow exchange with external water, we can have a first-approximate idea of the problem by diluting the discharge into the whole basin; C (t) concentration will then be

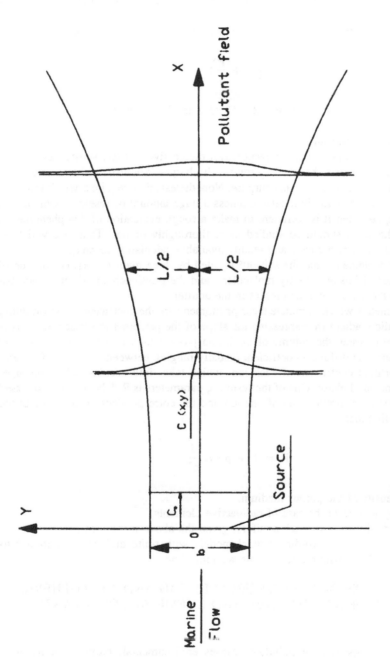

FIGURE 5.   Brooks' model: two-dimensional diagram of a polluting current in the sea.

$$C(t) = C_0\left(1 - \exp - \frac{P}{V}\,t\right)$$

where P is the emission rate of the contaminant, Co is the initial concentration, and V is the basin volume.

If the contaminant is radioactive, the argument of the exponential becomes:

$$-\left(\lambda + \frac{P}{V}\right)t$$

where λ is the decay constant, equal to $0.693/T1/2$ with $T1/2$ as half-life.

D. Diffusion Through Subsoil

The possible contamination of groundwater with radioactive or toxic substances is more likely to occur by accidents rather than by routine discharges. Sophisticated two- and three-dimensional models are available to this purpose. Nonetheless, they require a deep knowledge of the water table which is rarely available unless a huge amount of research is made. On the other hand, quite often it is sufficient to make a rough evaluation of this phenomenon to understand whether it should be studied more thoroughly or not. Thus, we will limit ourselves to touch upon rather easy and readily available calculation techniques.

Diffusion of contaminants into the subsoil[6] is ruled by three main factors: convection of underground water, diffusion (mainly molecular), and the physicochemical interaction between water and the solid materials present in the aquifer.

The main parameters which dominate these phenomena are the coefficient of permeability (m/s) and hydraulic gradient (h) expressing the slope of the piezometric surface, n porosity which is the ratio between the volume of the liquid phase which is not binded to the soil and the total volume, D diffusion coefficient (usually ranging between 10^{-5} and 10^{-8} cm²/s), and kd or partition coefficient of each ion, which takes into account solid exchanges between solute and solid phase. One of the resulting parameters is R delay factor, expressed as the ratio between the motion time of one ion and the correspondent motion time of the water table. We find that:

$$R = 1 + \rho \cdot K \frac{d}{n^2}$$

where ρ is the density of the porous medium.

Decay will play its role in the case of radioactive elements.

In order to study diffusion through aquifers, given the above-mentioned parameters, we usually adopt some one- or two-dimensional models with finite difference solution[7,8] to finally study the problem within a reasonable working time.

IV. EVALUATION OF EFFECTS FROM LIQUID AND ATMOSPHERIC DIFFUSION OF RADIOACTIVE AND CHEMICAL POLLUTANTS

A. General

The atmospheric dispersion of pollutants causes environmental, harmful, or negative effects on vegetation, animals, and man. In order to confine our analysis, we address our attention to man in particular. This limitation is more apparent than real; in fact, man is quite often one of the most sensitive organisms of the ecosystem; this means that if we protect man we shall protect the rest of the environment as well. The primary pathways of interaction and return to man of released pollutants are inhalation and intake. Secondary

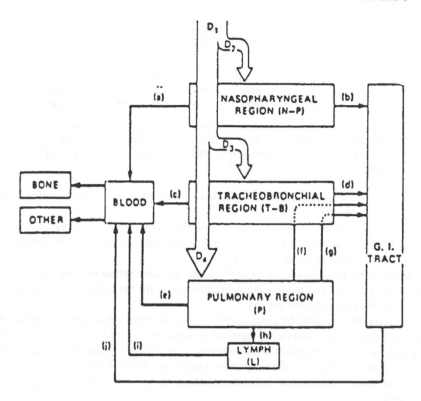

FIGURE 6. Compartment model for the inhalation pathway.

pathways are the surface contamination (skin absorption) and, for radionuclides only, external irradiation. Detailed analytical techniques for these cycles have been developed especially for radioactivity, but can be easily applied to other pollutants.

Since the ultimate purpose of this process is a risk evaluation for every plant dealing with toxic substances we should ideally know for each substance its effects on man, its degree of tolerance, and risk. However, only a few substances have been studied so deeply; in particular, modern chemistry daily "produces" new substances and we hardly know their acute toxic effects on experimental animals while few data are available for man. The only exception is given by radioisotopes which are subject to precise limits[9,10] that often take into account age of people at risk, as far as the inhalation and intake of these substances are concerned.[11] Precise limits (TLV) exist for some chemical substances and their prolonged or short exposure by workers.[12] From these limits we can extrapolate criteria valid for the general population as well, provided that a thorough bibliographic review of their experimental basis justifies the extrapolation.

B. Inhalation Risks

The inhalation of toxic or radioactive substances is the most probable risk, principally in case of accident, since breathing is a continued and fundamental function of the human body. As to radioactive risks, this function has been accurately modeled by an ICRP working group;[13] the human body was divided into eight sections (nasopharyngeal and tracheobronchial regions, lungs, blood system, lymphonodes, bones, gastrointestinal tract, and other organs) to study the pathway of the inhaled material through the various sections (Figure 6). We shall not go into detail but it should be noted that a large amount of the inhaled material is kept by the hairs of nasal and tracheobronchial tracts to be subsequently transferred

Table 2

AIR INHALED DAILY

	Liters of inspired air			
Activity (8 h)	Man adult	Women adult	Child (10 years)	Infant (1 year)
Light work	9600	9100	6240	2500 (10 h)
Nonworking activity	9600	9100	6240	1300 (14 h)
Resting	3600	2900	2300	3.8×10^3
Total	2.3×10^4	2.1×10^4	1.5×10^4	

to the gastrointestinal system by means of mucus; in this way, a new absorption path similar to that of food is created.

The same ICRP group[14] has prepared a table of inhaled volumes according to age and the type of human activity (Table 2). For quick evaluations, the mean breathing rate for adults can be assumed as equal to 1 m^3/h.

These considerations should be related to the air concentration obtained by methods described in Section II and the threshold concentrations for the various contaminants which are found in literature (see Section A).

Careful consideration shall be given to the different types of inhalation risk, either chronic in working environments and in the open air or acute in case of short accidental episodes.

As far as radioactive materials are concerned, it is possible to directly evaluate doses to individual organs and the whole body after inhalation, while in other cases we will limit ourselves to make a comparison between expected values and TVL, referring to short or long periods.

A special inhalation pathway which is often neglected is given by the resuspension of material deposited on ground or surfaces. To understand the magnitude of this phenomenon, we can say that in a "clean" air environment there are 0.1 to 1 mg/m^3 while visibility is drastically reduced when the dust concentration is around 1 g/m^3. Thus, given the average concentration in the original material (contaminated soil, dust) it is possible to estimate the maximum expected concentration present in the air. In order to assess the indoor air concentration, we shall consider that the concentration produced by an I source (kg/h) in V volume (m^3), with an R (h^{-1}) hourly exchange is given by the expression

$$C = \frac{I}{RV} \cdot (1 - e^{-0.693 \, Rt})$$

where t is expressed in hours. At the equilibrium,

$$C_i = \frac{I}{RV}$$

From the risk standpoint, we should make a distinction between contaminants which are directly felt by the human body (smell, color, and eye or mucous membrane reaction) and allow an immediate protection, and the wide range of substances, including radionuclides, which are not felt by our sense organs and thus require more or less sophisticated and timely measuring and alarm system.

Thus, radioactivity can be easily and readily monitored while for many chemical substances sophisticated and time-consuming lab techniques are required. It follows that one of the first priorities is the identification of the risk factor and its atmospheric diffusion effects.

In order to clarify the limits and meaning of such analysis we will make an example concerning the evaluation of the inhalation risk along with nuclear plant effects.

A nuclear power plant provided with good gas treatment equipment could discharge into the external environment 0.2 Ci of I^{131} during 1 month.

According to the meteorological analysis, the diffusion coefficient was 10^{-6} at a distance of 2 km, where a farm with cows and vegetables was present. The mean air concentration is then of 0.07 pCi/m^3 while the surface contamination is of 0.0002 $\mu Ci/m^2$.

Inhalation by infants entails an absorption of approximately 10 pCi of I^{131} corresponding to a thyroid dose of 0.16 mrem. Should the discharge go on throughout the year, we would have a dose of about 20 mrem, much less than the project value (30 mrem/year). Since this dose shall be added to the dose from ingestion pathway not yet known, we cannot give any final judgment as to the adequacy of the treatment system.

C. Ingestion Risks

At first sight, the risk from the intake of toxic or radioactive substances appears to refer essentially to chronic cases rather than accidents. However, if we take the Mihamata accident where consumers of seafood were severely and fatally affected by the concentration of mercury in mussels or the case of Chernobyl, it is clear that the intake pathway should be included in the analysis of toxic and radioactive substances environmental diffusion.

The path covered by the material from the source to man is longer and more tortuous than inhalation and presents subsequent processes of dilution and concentration not always well known.

The polluting source can be either airborne or liquid, since contamination may depend upon the atmospheric deposition onto ground and vegetables or the use of irrigation waters or even upon marine or freshwater fishing.

Among the other peculiar features of this risk we recall the prolonged stay of pollutant in the environment for long periods which could even cover many seasons, as well as the different water and fat affinity (substances soluble in water or fats).

As usual, a systematic analysis of pathways has been made for most fission and activation radionuclides of the uranium cycle in order to obtain sophisticated models and a huge series of environmental data; this technique is used to analyze the various problems linked to the intake risk.

The first pathway to consider is the liquid intake through water. If we take a daily water need for adults of 2 l, we can assume that 1 l is ingested with food and 1 l as drinking water. The reference concentration in water is evaluated as indicated in Section III. As far as radionuclides are concerned, there are some tables indicating both MAC (maximum admissible concentration) for workers and the general public,[15] and the maximum intake limit with the dose to the various organs.[11] For other substances, data on the toxic effects of the different industrial chemicals (e.g., heavy metals or pesticides)[16,17] are much more scattered and rare in literature.

The second intake pathway is represented by fruit and vegetables ingestion. They can absorb toxic substances in different ways: direct leaves contamination, dry deposition, irrigation and rainfall, and radical contamination of the soil. Higher concentrations are more often found in leaves than in roots, since the latter have strong physiological barriers capable of limiting the intake of contaminants. Concentration factors based on ionic concentration in the soil and in the vegetable are employed for the transfer of radioactive material. These values are approximate, since the actual intake depends upon the real ground concentration, pH, and the chemical structure of the material.

Table 3 shows a series of general bioaccumulation factors for vegetables, drawn from Reference 18 while Reference 19 provides more information on individual vegetables.

To make an intake estimate, we should define the average consumption of the various products through local studies or national statistics, while for the risk evaluation it is necessary to define the so-called "population critical groups". These groups include those individuals

Table 3
STABLE ELEMENTS TRANSFER DATA

Element	B_{iv}, vegetable/soil	F_m (cow), milk (d/l)	F_f, meat (d/kg)
H	4.8E 00	1.0E-02	1.2E-02
C	5.5E 00	1.2E-02	3.1E-02
Na	5.2E-02	4.0E-02	3.0E-02
P	1.1E 00	2.5E-02	4.6E-02
Cr	2.5E-04	2.2E-03	2.4E-03
Mn	2.9E-02	2.5E-04	8.0E-04
Fe	6.6E-04	1.2E-03	4.0E-02
Co	9.4E-03	1.0E-03	1.3E-02
Ni	1.9E-02	6.7E-03	5.3E-02
Cu	1.2E-01	1.4E-04	8.0E-03
Zn	4.0E-01	3.9E-02	3.0E-02
Rb	1.3E-01	3.0E-02	3.1E-02
Sr	1.7E-02	8.0E-04	6.0E-04
Y	2.6E-03	1.0E-05	4.6E-03
Zr	1.7E-04	5.0E-06	3.4E-02
Mb	9.4E-03	2.5E-03	2.8E-01
Mo	1.2E-01	7.5E-03	8.0E-03
Tc	2.5E-01	2.5E-02	4.0E-01
Ru	5.0E-02	1.0E-06	4.0E-01
Rh	1.3E 01	1.0E-02	1.5E-03
Ag	1.5E-01	5.0E-02	1.7E-02
Te	1.3E 00	1.0E-03	7.7E-02
I	2.0E-02	6.0E-03	2.9E-03
Cs	1.0E-02	1.2E-02	4.0E-03
Ba	5.0E-03	4.0E-04	3.2E-03
La	2.5E-03	5.0E-06	2.0E-04
Ce	2.5E-03	1.0E-04	1.2E-03
Pr	2.5E-03	5.0E-06	4.7E-03
Nd	2.4E-03	5.0E-06	3.3E-03
W	1.8E-02	5.0E-04	1.3E-03
Np	2.5E-03	5.0E-06	2.0E-04

who are more exposed to risk because of their higher consumption of certain agricultural products. For example, a person who buys salad in a shop is less exposed to risk than the farmer who uses his own products, if his farm is in the vicinity of the polluting source; moreover, the larger availability of a product (e.g., fruits) markedly increases its average consumption. A special case is that of fishermen in areas exposed to pollution risks. In fact, fishermen eat much seafood, far beyond the national average. In these cases, it is necessary to make an interview investigation of fishermen and their families to select a statistically significant sample. Individual consumption figures shall be treated by the cumulative frequency method in order to obtain median values and higher percentiles (90 to 95%). The set of data fits to a lognormal rather than Gaussian distribution.

Another factor influencing the ingestion of contaminant is the decontamination achieved through industrial processes of food cooking and meal preparation. The concentration may be reduced by a factor of 10, principally in case of leaf contamination.

The third intake pathway is represented by cattle and farm animals which are contaminated by watering or by feed and forage. Among these products we find meat, milk and cheese, and eggs.

Apart from the case of radioisotopes which is very well covered in literature, for other chemical substances any available information must be gathered taking into account the type (soluble in fats or not) and persistence in the environment. Also in this case, special attention must be paid to the critical groups and their diet; for milk, there is the special case of infants who depend almost exclusively on this food through the 1st year of life. Thus, it becomes critical in particular environmental conditions. As far as concentration factors are concerned, a major role is played by breeding techniques (open graze or use of industrial forage and feed), product, and derivate distribution markets (dairy industry, milk skimming, etc.).

Therefore, the first quantitative approach to the intake risk is based on the calculation of food contamination. To this end, we will give some useful indications, for accidental situations in particular.

In the case of atmospheric release, contamination of fruits and vegetables can be assessed through the following expression:

$$C_v = \frac{RGT}{Y}$$

where C_v is the vegetable concentration (kg/kg, Ci/kg), R is the fraction of contaminant kept by the vegetable, G is the contaminant fallout ($kg/m^2 - Ci/m^2$), T is the fraction remaining in the edible part, and Y is the vegetable productivity (kg/m^2).

Milk and meat contamination can be assessed as follows

$$Ca = C_v B_a U$$

where Ca = concentration in the animal product ($kg/kg - Ci/kg$), C_v = concentration in vegetables, B_a = bioaccumulation factor (d/kg), and U = daily quantity of ingested food (kg/day).

Table 3 shows B_a factors of milk and bovine meat, while for some nuclides and other zootechnical pathways refer to Reference 20.

The final important intake pathway includes seafood. Table 4 shows the concentration factor for marine and freshwater fish, crustacea, and shellfish of different elements including radionuclides.[18,21]

Such values clearly depend upon physiological animal needs and the amount of element present in water.

The high concentration factor of phosphorus, a scarcely soluble and extremely important element for life, should be noticed.

As to more complex chemical substances, we should refer to individual studies (e.g., see data on DDT concentration in cetacean fats) taking into account that in this case filtering organisms (mussels, bivalves) will probably accumulate foreign substances. Similarly to agricultural products, it should be possible to identify a critical group.

If we refer to the previously mentioned example, I^{131} vegetables contamination will be equal to 0.002 × 0.5 × 0.8/2 = 400 pCi/kg, where 0.5 is the fraction kept by the vegetable, 0.8 is the fraction left on leaves after washing, and 2 is the production expressed in kilograms per square meter. A consumption of 0.1 kg/d over 1 month corresponds to an intake of 1200 pCi, i.e., a dose of 32 mrem. Milk concentration can be assumed equal to 100 pCi/l. A 1-year-old child drinks 0.75 l/d, corresponding to 2250 pCi in 1 month, i.e., a dose of 60 mrem. The total monthly estimated intake dose is then 90 mrem. For the present standard, this dose is unacceptable and a drastic reduction of discharges should be required.

D. Other Risks

Apart from ingestion and inhalation, the environmental diffusion of contaminants entails other risks; many substances can be absorbed through skin, while radionuclides present in the air or on surfaces emit γ radiations, which are directly absorbed by man. For example, a first estimate of the irradiation risk due to a radioactive cloud is obtained through the

Table 4
TRANSFER FACTORS (pCi/kg/pCi/l)

Element	Freshwater		Saltwater	
	Fish	Invertebrate	Fish	Invertebrate
H	9.0E-01	9.0E-01	9.0E-01	9.3E-01
C	4.6E 03	9.1E 03	1.8E 03	1.4E 03
Na	1.0E 02	2.0E 02	6.7E-02	1.9E-01
P	1.0E 05	2.0E 04	2.9E 04	3.0E 04
Cp	2.0E 02	2.0E 03	4.0E 02	2.0E 03
Mn	4.0E 02	9.0E 04	5.5E 02	4.0E 02
Fe	1.0E 02	3.2E 03	3.0E 03	2.0E 04
Co	5.0E 01	2.0E 02	1.0E 02	1.0E 03
Ni	1.0E 02	1.0E 02	1.0E 02	2.5E 02
Cu	5.0E 01	4.0E 02	6.7E 02	1.7E 03
Zn	2.0E 02	1.0E 04	2.0E 03	5.0E 04
Br	4.2E 02	3.3E 02	1.5E-02	3.1E 00
Rb	2.0E 02	1.0E 03	8.3E 00	1.7E 01
Sr	3.0E 01	1.0E 02	2.0E 00	2.0E 01
Y	2.5E 01	1.0E 03	2.5E 01	1.0E 03
Zr	3.3E 00	6.7E 00	2.0E 02	8.0E 01
Nb	3.0E 04	1.02 02	3.0E 04	1.0E 02
Mo	1.0E 01	1.0E 01	1.0E 01	1.0E 01
Tc	1.5E 01	5.0E 00	1.0E 01	5.0E 01
Ru	1.0E 01	3.0E 02	3.0E 00	1.0E 03
Rh	1.0E 01	3.02 02	1.0E 01	2.0E 03
Te	4.0E 02	6.1E 03	1.0E 01	1.0E 02
I	1.5E 01	5.0E 00	1.0E 01	5.0E 01
Cs	2.0E 03	1.0E 03	4.0E 01	2.5E 01
Ba	4.02 00	2.0E 02	1.0E 01	1.0E 02
Ia	2.5E 01	1.0E 03	2.4E 01	1.0E 03
Ce	1.0E 00	1.0E 03	1.0E 01	6.0E 02
Pr	2.5E 01	1.0E 03	2.5E 01	1.0E 03
Nd	2.5E 01	1.0E 03	2.5E 01	1.0E 03
W	1.2E 03	1.0E 01	3.0E 01	3.0E 01
Np	1.0E 01	4.0E 02	1.0E 01	1.0E 01

following expression, which is applicable when the cloud size is infinite (i.e., higher than 100 m):

$$D = 0.25 \ EC$$

where D = dose intensity (rad/s), E = energy of γ emitted by radionuclide (Mev), and C = radionuclide air concentration (Ci/m^3).

A slightly more complex formula applies to the surface contamination irradiation.

It is then possible to evaluate the external irradiation dose of an individual by his time of permanence in the contaminated area. It should be noted, however, that in the night and part of the day people stay at home where they are partially shielded by external radiations. The reduction factor depends upon the type of houses (made either of stone, bricks or hollow tiles, wood, etc.), the number of windows, and the radiation energy. It is usually comprised between 0.15 and 0.3 according to the height of the apartment (basement and top floor).

V. CONCLUSIONS

From the above description it is easy to understand that the evaluation of the environmental diffusion of toxic and radioactive substances is a complex and multidisciplinary process.

Apart from technical industrial processes, it requires a deep environmental knowledge (meteorology, hydrology, socioeconomics, etc.) and a good sensitivity in the application of some often approximate methods and in the correct evaluation of results which are bound to be inevitably uncertain, as it has been stressed several times. This uncertainty should be kept in mind when making decisions on safety and protection of workers and the general public. Figures derived from this analysis shall be verified through field measurements, when possible. Furthermore, we should initially define the criteria which led to the selection of parameters; very conservative criteria might produce estimates far from reality, while a series of simplifications tends to overestimate risks.

However, the above-described procedure allows to systematically enlighten weak points, deficiencies, and the most common risk pathways; thus, we are compelled to study the environmental impact of industrial processes. It is surely better to prevent effects and consequences rather than bitterly acknowledge them.

Finally, it should be noted that a correct evaluation of the environmental diffusion of dangerous materials and of its effects often becomes a true saving source, in terms of time and money and in spite of its costs; suffice it here to say that a containment system for the "vessel" blow down from the safety rupture disk of Seveso costs some million lira, while the accident consequences cost at least as one thousand times as much.

REFERENCES

1. **Slade, D. H.,** *Meteorology and Atomic Energy,* 1968.
2. **Martin, J. R.,** Recommended Guide for the Prediction of the Dispersion of Airborne Effluents, 3rd ed., ASME, 1979.
3. Atmospheric Dispersion in Nuclear Power Plant Siting, IAEA Safety Series No. 50-SG-S3, 1980.
4. **Cagnetti, P. and Ferrara, V.,** Two possible simplified diffusion models for very low wind-speed, *Rev. Mat. Aeron.,* V XLII nr 4—1982.
5. **Pearson, E. A.,** *Waste Disposal in the Marine Environment,* No. 5, Pergamon Press, Elmsford, NY, 1960.
6. **Till, J. E. and Meyr, R. H.,** Radiological Assessment, NUREG/CR/3332.
7. **Niels, I. J. and Carlsen, L., P.Bo,** Column 2. A computer program for simulation of migration, Riso National Laboratory, 1984.
8. **Mancini, O. and Gera Miracq, F.,** A Program to Simulate Radionuclides Migration in Porous Media, Developed by ISMES within the framework of ENEL contract no. 3369, 1985.
9. ICRP, *Limits for intakes of radionuclides by workers,* Publ. No. 30.
10. **Keveling Buisman, A. S.,** From Body Burden to Effective Dose Equivalent. Parts I and II, Report ECN 125, Netherlands Energy Research Foundation.
11. **Breur, F., Brofferio, C., and Sacripanti, A.,** Committed Dose Equivalents due to the Intake of Radioelements by Four Age Categories, ENEA/RT/PROT, (83) 24, 1983.
12. ACGIH, TVLS, Threshold Limit Values for Chemical Substances in Workroom Air, Adopted by ACGIH for 1976.
13. ICRP Task Group on Lung Dynamics, Deposition and retention models for internal dosimetry of the human respiratory tract, *Health Phys.,* 12, 173, 1966.
14. Task Group Report on Reference Man, ICRP Publ. 23, 1975.
15. Council directive of July 1980 amending the directives laying down the basic safety standards for the health protection of the general public and workers against the dangers of ionizing radiation, CEE, EUR 7330, 1981.
16. FAO/ONS, Liste des concentrations maximales de contaminants recommandées par la Commission Mixte FAO/ONS su Codex Alimentarius, CAC/FAL 3, 1976.
17. *International Regulatory Aspects for Pesticides Chemicals,* Vettorazzi, G., Ed., CRC Press, Boca Raton, FL, Vol. 1, 1979.

18. Calculation of Annual Doses to Man from Routine Releases of Reactor Effluents for the Purpose of Evaluating Compliance with 10 CFR 50 App. I, Regulatory Guide 1.109, Revision 1, U.S. National Regulatory Commission, Washington, D.C., October 1977.
19. Ng, Y. C., A Review of Transfer Factors for Assessing the Dose from Radionuclides in Agricultural Products, Nuclear Safety Vol. 23, No. 2, January-February 1982.
20. Boone, F. W., Ng, Y. N., and Palms, J. M., Terrestrial Pathways of Radionuclide Particulates, Vol. 41, 1981.
21. Blaybock, B. G., Radionuclide Data Bases Available for Bioaccumulation Factors for Freshwater Bioo, Nuclear Safety Vol. 23, No. 4, July-August 1983.

Chapter 11

EXPOSURE LIMITATION IN NORMAL OPERATING CONDITIONS

L. Bramati

TABLE OF CONTENTS

I. INTRODUCTION

This section deals with chronic risks both for workers and the general public. These risks are linked to the normal operating conditions and include both malfunctions and maintenance, while we shall not take into account risks due to accidents for which preventive and safety methods exist. Since risk exposure and protective measures are completely different for workers inside the plant and people in the vicinity, we have separated the two problems.

Also, exposure limits are different for the two categories: while workers are exposed only 8 h/d and plant controls are stricter, the population is exposed 24 h/d and "spot" controls are not even carried out in all cases.

As we have said in the previous chapter, toxic or radioactive risks for man come from inhalation, ingestion, skin contact, and external irradiation as far as radioactivity is concerned. Thus, the limitation of exposure is based on project and actual operating conditions to minimize the environmental risk and protect every individual by means of training, protective means, and behavior rules. It might seem obvious, but the risk limitation procedure is based on the correct and complete identification of risks at the plant. This identification is not always evident as one could believe. We must know productive processes in all their aspects and have a list including all substances used, final products, as well as expected and even undesired byproducts. A classical example showing the problem complexity is the Seveso accident; dioxin (TCDD) was only the final residual substance of a procedure in which reaction temperature played a major role but no toxic substance had been originally employed.

Knowing the type of materials and procedures employed, each of them should be assigned a predicted risk factor. Tables describing limits, symptoms, and effects are available for the most important substances including radioactive substances, while for others only a small amount of information, if any, is available unless one makes an in-depth literature search.

Only at the end of this procedure will it be possible to study some methods for limiting the exposure by workers and the population.

II. LIMITATION OF CHRONIC RISKS FOR PLANT WORKERS

A. Project Provisions

Some project characteristics (layout, ventilation systems, monitoring systems) allow to reduce occupational chronic risks due to toxic and radioactive materials while others are especially referred to radioactive risks (shields, equipment segregation, choice of materials). The location of premises and equipment of those industries which process dangerous materials should minimize undue exposure risks during the actual operation and maintenance of the plant. This means that corridors and access pathways should reduce the risk residence time, there should be wide drainages without dead points, floors and walls should be easily washable (avoiding cutting edges, using paints and tiles easy to decontaminate), while some areas where to safely carry out maintenance operations should be provided. Experience has shown that during the project development phases there is the tendency to concentrate equipment, pipes, and cables all together, resulting in a final overreduction of space that may create problems of access and interferences even during normal operations. Apart from industrial hygienic considerations, space saving is often translated into a loss of time and organizational problems during work and maintenance.

Plants often have only a limited number of processes or equipment which might be an occupational risk source. In order to limit dangers and reduce the number of workers exposed, premises and services should be segregated or classified.

Nowadays, plants are full of control rooms. The location and design of the control room should take into account the presence of dangerous materials or of the operations that could

Table 1
SHIELDING THICKNESS
CORRESPONDING TO AN
ATTENUATION OF 1/10 (cm)

Material	γ-ray energy (Mev)			
	0.5	1	2	3
Lead	1.5	3.3	5	5.4
Iron	5.5	7	10	11
Concrete	19	23	30	33
Water	60	66	82	90

originate them. Risk phases should be clearly indicated while a good lighting and alarm system should be provided. Since workers might be contaminated during normal activities, the plant must be provided with changing rooms, toilets, and showers in the immediate vicinity of risk areas. If plants are to be maintained by a large external staff, changing and decontamination rooms including provisional facilities (cabinets, showers, etc.) shall be provided for additional people.

In plants dealing with dangerous materials, the major practical risk for workers is represented by the inhalation of gases, dust, and fumes.

It is extremely important to have an efficient ventilation system where the air flow from safer areas is conveyed towards areas exposed to higher risks, while the most dangerous operations should take place under suitable ventilated hoods. Air should be changed very effectively (two to four air changes per hour) while special filters shall be installed near the operating areas during maintenance.

In order to apply these procedures and make use of the protective means listed below, it is necessary to continuously evaluate the risk level for the various parts of the plant; while some contaminants provoke an immediate response by the human body, others (most of them) are not felt at all, especially in case of chronic contamination. For example, radioisotopes cannot be identified. For this reason, it is necessary to have a monitoring system to identify the ambient contamination (gas, aerosol, or radiations) especially in high-risk rooms.

At this point, we should consider the design features typical of nuclear plants: shields and structural materials. Radioactive sources in nuclear plants must be protected to minimize both the environmental diffusion and the direct irradiation risk. To this special purpose, it is necessary to install fixed (concrete, lead, iron) or mobile (shielded doors, mobile panels, etc.) shields. Shielding engineering is quite sophisticated and is based on precise and complex calculations[1] giving the expected radiation level with 30 to 40% error rate. A simpler method takes into account shield thickness values reducing the radiation intensity by 10 factor, as those shown in Table 1.

It is advisable to segregate the most radioactive equipment to reduce the exposure of operating and maintenance staff. When the exposure outside such premises is considered "undue", it is necessary to pass across shields or labyrinths before going in. As far as radioprotection is concerned, the "ALARA" (as low as reasonably achievable) concept is applied;[2] according to it, the risk decrease must be offset by a corresponding social or economic cost. For instance, it is useless to increase the source shielding for an adjacent room where nobody enters or where the radiation level is measurable; on the other hand, even a "low" dose level is unacceptable if one can further reduce it by means of simple and inexpensive means.

Another provision to limit occupational exposure in nuclear plants concerns the selection of structural materials. In the last 10 years, fuel technological developments have almost eliminated chronic leaks of fission products. Thus, it has been demonstrated that the workers' collective dose is almost exclusively due to irradiation from crud in the primary fluid which deposit on equipment and pipes. In particular, ^{60}Co, a radioisotope resulting from the irradiation of stable cobalt and present in very small quantities both in steel and nickel, plays a major role in this process. Thus, efforts have been made to reduce the amount of this element in inconel and stainless steel and replace cobalt alloys (stellites) with other similar materials in valves and abrading parts. This initial choice leads to a dose decrease in the following years and does not require large economic investments.

B. Operating Provisions

In order to limit the occupational chronic exposure it is necessary to correctly operate the plant after designing it according to criteria listed in Section A. This means that plant managers should be really aware of the necessity of limiting chronic risks by adopting precise rules.

The most fundamental means to be used to this purpose are a "dedicated" unit, oriented operating procedures, availability of protective means, and personnel training.

Nuclear power plants, traditionally exposed to certain risks, can be taken as an example of the application of the above-mentioned criteria. However, these criteria can be applied also to other plants involving occupational chronic risks. Differently from what one could superficially believe, the firm application of those criteria may contribute to reduce the number and duration of interventions both in operating and maintaining conditions. In the modern industrial world, where our society is increasingly aware of environmental risks, industrial hygiene actually brings its own contribution and offers long-term economic advantages. We will try to develop the above-said concepts analyzing the first five factors which may limit both chronic and accidental risks.

First, it should be noted that plants involving occupational risks are industrial plants that aim at obtaining the highest return on investments. Thus, the operating and maintenance staff will hardly be in charge of risk-limiting activities since they would probably tend to prefer other tasks, should a conflict of interest among the various needs exist.

In nuclear plants, it is the radioprotection unit to be in charge of risk limitation. This unit may even account for 5 to 10% of the operating staff. In other types of plants the same team is called "industrial hygiene unit" and its tasks change according to the amount of risk; however, two points should always be observed:

1.   The staff in charge of the industrial hygiene of the operating and maintenance line shall be independent.
2.   They should always have the possibility to express their opinion on plant activities (consulting activity).

This second point is linked to the need of requiring the opinion of the industrial hygiene unit in all cases. The concept especially applies to maintenance services which are not carried out routinely but may equally entail high occupational risks. In nuclear plants the procedure is the following: the "radioprotection unit" signs a "working permission" whereby they indicate the probable risks that will be encountered along with the measures that should be adopted. Of the routine work may by the industrial hygiene unit we recall the recording of the atmospheric and surface contamination at work sites and the review of those procedures which form the basis of their consulting activity.

As for the operating procedures which limit the occupational exposure, other ways to either directly or incidentally limit risks (shifts relays, operating limitations to operating parameters, etc.) may be included. As far as nuclear plants are concerned, it has been observed that a basic factor to limit both the dosage and occupational exposure is the wise management of the primary water chemistry to decrease corrosion and deposition of crud. Regarding high-risk contamination areas, individual protective means shall be provided. The employment of clothes, masks, and self-contained breathing apparatuses is burdensome but quite indispensable. The industrial hygiene unit shall be in charge of these means and will take care of their distribution and collection and supply and storage, controlling their functionality and adequacy. Nuclear plants are provided with an "active laundry" for the daily decontamination of the objects which are commonly used in controlled areas, above all during maintenance periods. Careful consideration should be given to this kind of activity and structure to prevent it from becoming the "bottleneck" of the system, with severe consequences for industrial discipline and safety as well as for the implementation of normal activities.

The last chronic risk-limiting measure consists of a correct training of the personnel. Every member of the staff should know risks, protective measures, and procedure enforced. It is this activity which demonstrates the managers' will to effectively reduce risks. In fact, instead of carrying on their normal tasks, the staff will take part in training and refreshment courses at regular intervals. This "distraction" is often reluctantly accepted by department managers and by the same staff members.

III. LIMITATION OF CHRONIC RISKS OUTSIDE THE PLANT

A. General

As already discussed in Section I, the general public and people living in the vicinity of plants are exposed to some risk, due to the environmental diffusion of pollutants which might return to man through the so-called "critical pathways". In order to limit risks it is necessary to go through three steps: to reduce the amount of substances to the environment by means of special treatment facilities, to have an in-depth knowledge of pollutants and their effects on the biological physical-chemical chain connecting plant and man, and finally to make environmental controls especially in the most exposed or "critical" sectors according to the environmental studies.

B. Plant Provisions for the Limitation of Risk for the Population

After carefully analyzing the plant design and its operating methods, it is possible to identify the diffusive pathways of dangerous materials to the external environment. These are usually represented by stacks, ventilation towers, sewer systems, and sewage and solid waste discharges. In this case, it would be advisable to check the amount of releases and eventually employ special systems for their elimination. However, before employing these burdensome and expensive facilities we will carry on the environmental analysis discussed in the previous chapter as a starting point.

After studying discharges, corrective measures, and environmental effects it will be possible to achieve an optimal solution.

In fact, very often it is advisable to adopt an easily manageable treatment system with low decontamination factor rather than a highly sophisticated system that will never work because of operating difficulties and the lack of motivation in terms of negative effects both for the environment and the populations; in other words, one should avoid an "overkilling" situation.

Airborne effluents "controlling" techniques range from the simple filtraton to absolute and active charcoal filters. Large plants often make use of electrostatic precipitators, bag houses, and scrubbers. Beside the treatment plant design, it is necessary to decide how to dispose of the material remaining at the end of the treatment by recycling it within the same plant, disposing it in the environment or storing it locally.

Liquid effluents are generally stored in the head tanks of the purification system. They can consist of settling ponds, flocculators, filters and centrifuges, ionic exchange columns, and evaporators. The order shown reflects growing difficulties and contamination factors. The resulting solid waste disposal constitutes a problem for all kinds of plant. Both production lines and effluent treatment systems generally produce solid wastes contaminated by process products and byproducts. The first way to diminish environmental effects consists of decreasing the amount of this waste, which should be carefully segregated also taking into account the contaminant concentration; it is wiser to store 10 high concentration drums rather than 1000 low concentration drums for which the plant does not have room enough. Solid waste can be either stored in bulk or in drums, tanks, and special containers. When the waste disposal occurs outside the plant, consideration should be given to their interaction with the environment (leaching by atmospheric agents, violation, dispersion, reutilization). The most dangerous waste is usually incorporated into low leaching materials (cement, bitumen, plastic material) to be segregated in special controlled discharges.

However, for all effluents it is necesary to define a limit below which the material, either toxic or radioactive, can be released without any further measure. Regulations do not establish these limits, since they result from the environmental analysis which takes into account the processes of dilution, reconcentration, and return to man. Such limits constitute the threshold limit of the treatment systems that is taken into account by the effluent monitoring system before the environmental emission. The monitoring system may be both a sample taken at the discharge point to be later studied in the laboratory and sophisticated continued measurements at all the discharge pathways of nuclear plants. However, for all risk plants the monitoring system should be defined by the project regardless of its complexity (sophisticated electronic equipment or simple procedures). The industrial hygiene or radioprotection unit (in nuclear plants) is usually in charge of its management.

C. Environmental Impact Evaluation

The evaluation of the plant environmental impact is the most important means to quantify the actual risk to which the population is exposed. As already widely described in Section I, this assessment requires a more or less profound knowledge of the whole process from the source to food and man, through air, water, and ground diffusion. It should identify critical pathways and elements as well as critical population groups.

The degree of knowledge required is related to the amount of risk. The risk evaluation starts from the approximate and strongly securing data to get a final detailed assessment of the environmental receptivity of the various materials. The term receptivity means the quantity introduced into the environment which produces a previously established health and environmental effect.

A practical example would serve to clarify the meaning of receptivity. The yearly discharge of 1 Ci of Cs^{137} into a river entails a partial return to man through the ingestion of irrigated fish and vegetables. Further, people who go fishing along the river will be directly irradiated by water and sediments on the banks of stagnation areas. The most critical group (farmers who also practice sport fishing) receives a total annual dose of 1.5 mrem.

As to the maximum dose admited for particular population groups (500 mrem/year), Cs^{137} receptivity in this case is equal to 330 Ci. As far as nuclear power plants are concerned, environmental impact evaluation (EIE) is used to prepare safety reports (both preliminary and final); it includes information to evaluate the effects from the normal plant operation as well as data concerning exceptional events (earthquakes, floods) and accidents. We will list the table of contents of the reports as it is made in Italy; it could be used as an example for other kinds of plant:

1.   Plant and site description
2.   Meteorology

3.   Hydrology
4.   Geology, seismology, and geotechnics
5.   Ecology
6.   Ground and water functions
7   Demographics
8.   Environmental radioactivity
9.   Environmental impact from radioactive, thermal, and chemical discharges
10. Complementary studies
11. Control networks
12. Effects of accidental events

It is obvious that listings 9 and 12 make use of all data collected in the previous chapters. Listing 10 points to the complementary investigation areas to be developed, while the contents of listing 11 — surveillance or control network — is partly discussed in paragraph Section D.

In 1985, the European Economic Community (EEC) issued a directive on the preparation of a sort of EIE for large plants which shall be enforced by member states. It concerns both high-risk plants and other projects of public interest, according to a special list attached to the directive.

The EIE contents is specified in Reference 5 of the directive. It comprises the description of the plant project, which should specify the type and amount of materials produced and discharged, the analysis of possible alternatives to the same plant, the description of environmental components involved, and a final evaluation of predictable effects and the ways to diminish them. The EEC EIE is very similar to that of nuclear plants safety report.

D. Environmental Control

When the plant impact evaluation stresses the presence of a possible risk for the population, environmental control measures to record the actual environmental concentration of materials discharged and their possible effects on both man and the ecosystem must be arranged. The network structure is defined on the basis of the environmental knowledge acquired (see section C) and VIA indications on critical pathways. According to the importance of the risk, the network consists either of some control points to be made at different times or of a series of sampling and measurement permanent stations. Environmental matrices to keep under control are usually air, water, ground, fish, and agricultural products although the choice depends upon the type of material and the discharge ways. Nuclear plants have usually a network consisting of some permanent stations which examine air and water samples, a number of points for the seasonal examination of agricultural products, and a dosimetric network as well. As far as river or sea discharges are concerned, also sediments and fish products are analyzed.

These investigation campaigns lead to an in-depth knowledge of the ecosystem segments as a routine network could never do. In fact, it should be said that apart from particular cases, it is rare to find a high contamination level in collected samples, thanks to the role played by the environmental diffusion. Thus, only very sophisticated lab techniques and a highly specialized staff can find the smallest environmental signs left by the plant. Measurement campaigns are intended to employ the best equipment and techniques available to get a high-definition picture of the ecosystem section defined as critical by the environmental analysis, with low cost in terms of time and financial resources. On the other hand, when positive results are found, it is possible to discover where they are from and finally reduce the population exposure resulting from the normal plant operation.

REFERENCES

1. **Jaeger, R. G.,** Engineering Compendium on Radiation Shielding, 1968.
2. Recommendation of International Commission on Radiological Protection, ICRP Publication no. 26, 1977.
3. Information Relevant to Ensuring that Occupational Radiation Exposure at NPP will be as Low as is Reasonably Achievable, Regulatory Guide 8.8, U.S. National Regulatory Commission, Washington, D. C., June 1978.
4. ENEA, Techanical guide no. 1. Recommandation. Documentation contents: a) preliminary project; b) preliminary safety report, March 1975.
5. EEC, Council directive dated 22 June 1985 on the evaluation of the environmental impact of certain public and private projects, Official Bulletin of the European Community, NL.175/40, European Economic Community, July 5, 1985.

Chapter 12

EXPOSURE LIMITATION FOLLOWING ACCIDENTS: SITING AND EMERGENCY PLANNING

Giancarlo Tenaglia

TABLE OF CONTENTS

I. INTRODUCTION

The main guarantee of limitation of the possible consequences of the operation of an hazardous plant in relation to the operations themselves, to the plant itself, to those who work there, and to the territory in which the operation takes place, must always be provided by the "intrinsic" safety of the plant, i.e., safety planned and defined in the design stage and subsequently implemented in the construction and operating stages. Taking account of a predictable, even though generic, territorial context, within which the plant must operate, this safety ought to be the expression of the safety level which it is desired to confer to the plant as such. It is also the standard of reference for any other protection of the plant which the specific operating context may require.

A further guarantee of the limitation of possible adverse consequences ought in all cases to be obtained through careful siting of the plant and through the preparation of emergency plans.

That said, there follows a discussion of the criteria and requirements which should govern the siting of potentially dangerous plants and the preparation of on-site and off-site emergency plans.

II. SITING

The term siting is usually used to mean the process by which a suitable site is located for a specific plant. In this process the essential characteristics of the plant must be known, while the site and its territorial context are the subjects of study.

The terms of reference for this process are the interactions which inevitably occur between the plant and the territory which hosts it during the various stages of the life of the plant from the beginning of its construction.

There are very many of these interactions and they differ greatly from each other. In essence they depend, on the one hand, on the specific type and scale of the activity which the plant carries out, especially on the territorial requirements of the plant and the various types of impact it may have on the surrounding environment. On the other hand, they depend on the characteristics of the territorial context in which the plant is sited.

The quality, quantity, and importance of these interactions determine the greater or lesser complexity of the siting process.

An important factor in this connection is the size of the area such interactions affect (in particular in the case of hazardous plants, the size of the area of potential impact). The larger this is, the larger the area to be studied, the more the aspects involved, the greater, other things being equal, the complexity of the impacts, the more difficult it is to find areas with characteristics such to permit drastic reduction, if not elimination, of adverse impacts, especially those in the immediate vicinity of the plant. A plant whose impacts may extend for at most 1 or 2 km is one thing; one whose impacts may extend for tens of kilometers or more is quite another. For example, the first situation is obviously a great deal easier if restricted areas have to be imposed around the plant.

However, it should be noted that in the process of selecting a specific site for a plant the effects which the plant may produce or undergo at close and medium distances are decisive. Effects which the plant may cause at great distances, let us say at hundreds of kilometers or more, if any, have little or no influence on a specific site, since this can normally be changed by several kilometers without significantly modifying long distance effects. In general, only a different decision in terms of large areas or geographical regions (1,000, 10,000, or more km^2) is significant for such effects.

The duration of interactions is also important in plant siting (usually they are not limited to the operating life of the plant, but begin with its construction and can continue even after final shut down as long as the site is not returned to its previous use), as is the reversibility of the impacts (e.g., permanent occupation of the site if the plant is not expected to be dismantled at the end of its life).

Finally, the nature of the implications of these interactions is very important. They are never exclusively technical and economic, still less so in the siting of major hazards plants where the size of some installations coupled with the potential dangerousness of some substances within them could mean that the consequences of an accident would be felt over such a large area that entire communities, not just small groups and limited economic interests, would be affected, and where the full social dimension must therefore be taken into account. True, the probability of such events occurring is normally very low, but the damage they cause is such that they must nonetheless be considered. In these cases the question of siting goes beyond not only the exclusively technical and economic (simple identification of feasible or economic sites) but also the private decision-making context of the industry concerned. It is a question the solution to which involves a number of actors with differing roles and responsibilities (those responsible for land management, industry, regulatory authorities, the inhabitants, etc.) and the decision-making process requires checking and bringing together at different levels in series of steps. Potentially opposed interests (e.g., safety of the public and economic production) must be reconciled in an integrated view of all the implications. The result, at times, is an extremely long decision-making process which ends by becoming a significant cost in both economic and social terms. One needs to think only about the consequences of delays in implementing large investments and about the involvement of the local population in expressing an opinion on the siting proposed. Even if current practice can and should be improved in many respects, it remains true that the siting of this type of plant is a fundamental component in many decisions connected with development policy.

It is not our intention to look in detail at these last points but rather at the implications which siting has on design for major hazards industrial plant. What follows will therefore be limited to a discussion of those aspects of the problem of siting which are primarily and directly connected with safety, well knowing that the distinction between safety and nonsafety aspects cannot be absolute and that at times nonsafety aspects impose special design requirements that may have an effect on plant safety features and that siting is not decided solely on considerations of safety but also of all the other aspects (technical, economic, social, cultural, etc.) which may also play a decisive role in the choice of a site.

From here on we shall refer only to the widest process of siting which envisages the possibility of choosing between alternative (or potential) options, whereas what is often asked for is only a check on the suitability of a site proposed for a given plant. The part of the process which, through studies, estimates and evaluations lead to ascertain the suitability of a site in comparison with the siting objectives exists in every case; what may be lacking in such cases is the optimization of the choice.

A. Siting Objectives

Referring specifically to major hazards plants and safety aspects, the most important siting objectives can be summarized as follows:

1. To make available everything required for a smooth and an economic construction and operation of the plant.
2. To protect the public and the environment from any adverse impact from the plant.
3. To make it possible to protect the plant from adverse effects induced by events external to it which might occur in the area of the specific site in question.

The first objective relates to the primary purpose for which the plant is constructed, the carrying out of a given activity. It should be emphasized that the availability referred to is not generic but interlinked with an overall evaluation of the cost and smoothness of plant construction and operation. Speaking very generally, what is required is availability of an area sufficient to contain the plant, service areas, and perhaps a larger or smaller restricted access area; soil with characteristics suitable for the plant foundations; power; water; road; and rail and sea connections for the transport of construction materials, raw materials required for operation, finished products, etc. The type of plant and its size will determine the amount and quality of the above requirements.

The second objective refers primarily to normal and accidental releases of polluting or dangerous substances and to instantaneous releases of great amounts of energy (e.g., explosions). However, it also includes protection of the environment from the adverse impacts which may occur during plant construction and protection of the public and of the environment from the activity carried out, particularly from its social and economic effects and from those on the habitat. The term "environment" must, in fact, here be interpreted in its widest sense: the natural, agricultural, and industrial environment and the human habitat.

The third objective is closely related to the first two. In fact it is pursued in the light of these. The events to which reference is made may be natural (such as earthquakes, floods, etc.) or human activities (such as explosions, falling aircraft, etc.). It is necessary to check whether adequate protection can be obtained by action on the plant design and if so at what cost, or simply, in the cases of human activities only, by adopting a safety distance from the site of the event.

B. Reference Criteria

To achieve the objectives stated above, it is necessary to have, or to establish, reference criteria with which to compare the results of territorial analyses and studies and estimates of the impacts which may follow installation of the plant.

Dealing with the subject in outline, it can be said that normally these reference criteria are expressed either as a threshold or as small number of continuous ranges. The threshold divides the possible values for a parameter into two fields in which the value judgment is precisely opposite (acceptable, examinable, permitted, recommended, below the threshold value and unacceptable, to be rejected, not permitted, not recommended, above it). For example, in checking the ability of a body of water to satisfy cooling requirements of a plant, the reference criteria can be expressed as the minimum flow recorded in the last 40 years, fixing a value for it of, for example, 10 m³/s to ensure full operation of the plant for 365 d/year.

In contrast, contiguous ranges graduate judgment of the quality or worth of values of a parameter within the range of acceptability. For example, there may be a division into three categories: the first one refers to an ideal situation; the second one to an acceptable situation on condition that other guarantees are provided; and the third one to a situation acceptable only with difficulty but not to be rejected a priori. For example, to evaluate distances of a site from sizable urban settlements, of the order of some tens of thousands of inhabitants, there might be three ranges in the reference criteria: distances of, say, over 15 km, distances between 10 and 15 km, and distances between 5 and 10 km.

Different criteria are used to establish values for thresholds and ranges, but in all cases they must be commensurate with the problem to be solved and the flexibility or otherwise one may have in finding solutions.

With thresholds, the process is clearly a yes-no one. As such it is quicker (a situation is either accepted or rejected and the analysis proceeds), more measurable, and in this sense it could be considered as more objective. However, it is also more rigid and may be more costly (situations may be rejected which would be acceptable in a deeper analysis: the threshold cannot but be selected for average rather than special situations). With ranges as reference criteria the process is more flexible (it permits graduating a judgment and therefore leaves more options open for the final stage of overall optimization of the various parameters which play a role in any decison), but is is also more complex (more factors are kept under review right up to the end of the process). However, the choice between the two processes is not always a free one.

Regulations relating to safety, pollution, land use, etc. tend to be expressed in threshold values, in order to facilitate application and checking. The maximum concentraton of a given pollutant in the air must not exceed a given value. The amount of oxygen in a water body must be above so much. The individual risk in certain conditions must not be greater than so much.

Regulations on the siting of industrial plants include both thresholds and ranges. Use of one or the other approach clearly depends on the context in which one is operating. In relatively simple situations and those in which siting parameters are not very interdependent, the selection process can advantageously be taken to the decision-making stage adopting fixed reference values for every parameter. On the other hand, when the parameters are markedly interdependent, the decision-making stage cannot but follow a process of optimization of the whole, each with its own weight in the decision. To achieve this, the parameters must be evaluated in terms of their numerical value or of a range, wide or narrow, of values into which they fall. In such cases one must therefore exclude, at least in the final stages of the decision, any approach with reference thresholds, though these might usefully be used in a preselection stage or in identification of areas within which a site might be found.

If there are safety issues with potenially important effects on the territory (and this is the situation we are referring to), there is another reason for adopting a process of optimization in addition to the fact that the parameters of importance for siting necessarily become interdependent. Safety can never be absolute, and even when thresholds of acceptability are established there is an obligation to reduce effects in the specific use within an overall optimization of all the parameters concerned. Equally, the more significant the territorial implications of a siting are, the wider must be the viewpoint, the decision-making context, within which optimization takes place. In other words, it must be less "sectoral" and less "local". Nor forgetting that the point of view which we are examining the siting of a plant is that of its influence on design, we should make it clear that having a general design for a plant to be sited means having already taken account of the range of variation of territorial parameters at the national or regional level (according to where the choice must be made) and having established a design for the plant to fit the siting objectives in at least some area of the country or region. Most of the design parameters will be set in this way, but others are left within a field of variation, to be defined on siting and in relation to the site characteristics. One might think, for example, of the question of the foundations or of the need to adopt some protection measures additional to those included in the general design either to keep some impacts within preset thresholds or to optimize others. These design parameters, which we can call very broadly site dependent, are those which take their place in the siting parameter context within which there is a possibility of optimizing a selection.

It is obvious that the selection of a site means not only the selection of a specific site but also of a context within which the site is located; a context which has been analyzed and evaluated within the framework of a range of possible variation for some of its parameters during the life of the plant (e.g., residential or industrial development or the construction of communications systems). Outside this context, the validity of the choice might be called into question. For example, the importance of some impacts, estimates of which played an essential part in the choice, might change significantly. Hence there is a possibility that the competent authorities may, on licensing the siting, impose restrictions or conditions on the development of the area in order to prevent residential settlements of certain size appearing within a given distance from the plant or else they may require a check of compatibility with the previous plant to be made before any new hazardous industrial establishment is erected within x kilometers of the plant in question.

It also goes without saying that the context within which detailed design must take place is determined together with siting. Consequently, every siting decison is accompanied by definition of basic site-dependent design parameters.

At this point we must clarify the question of who, in practice, sets or ought to set the required specific reference criteria for the siting of major hazards industrial plants. Still speaking very generally, we can say that the licensee prepares those reference criteria for siting which relates to smooth and economic construction and operation, while the public authorities prepare and issue recommendations and instructions on all matters affecting the public domain, the community, and especially protection of the public and of the environment. Some of these do not concern the siting of a specific hazardous plant directly, but the need to comply with them affects siting (e.g., zoning regulations or general pollution regulations). Those which affect siting directly are specific to the type of plant and closely linked to its intrinsic safety. Sometimes such regulations do not limit themselves to lay down thresholds or ranges of "acceptability" for some parameters of importance for siting, but also specify the process for identification and estimating these parameters and the methodologies to be used in estimating impacts.

Referring specifically to the regulations issued by public authorities, one can say that national and international standards for the siting of nuclear plant are well developed and also very detailed. The same cannot always be said for other types of major hazards plant. Indeed, the regulations are often very defective and the siting criteria then end up as, jointly or severally, existing practice for similar plant, the individual risk for those living closest to the plant, safety distances from other industrial plants, and so on. However, these are only temporary in the absence of a clear and public definition of their application on the one hand, and on the other of the present more active review of the safety criteria to which these plants have hitherto been designed, installed, and operated. But if it is not possible to give complete and consolidated reference criteria in view of the rapid evolution in safety standards, it is expected that it will shortly be at least regulated.

C. Selection Process

The aim of the process is to identify the most suitable site for the installation of a given plant within the region of interest. The region of interest may be a whole country, an administrative subdivision, or an area defined specifically (e.g., for technological, social, or economic reasons). In any case the region will be sufficiently large to give sense to the selection process.

Selection processes are not all of the same type. In fact they may vary considerably from case to case, depending on the characteristics of the plant and of the region of interest. Speaking very generally, however, it can be said that when the extent and complexity of the area studies required for the siting of a plant are great and/or when the number of possible sites available is also great, it is often advisable to make a preliminary survey of suitable areas in order to reduce costs (not only in time but also in money). The preliminary survey is carried out by initially examining a limited number of territorial parameters at the most preliminary possible information level, thus eliminating those areas which do not seem suitable on this initial examination and then selecting the best of the remainder for the subsequent studies to determine whether they contain one or more sites suitable for the plant in question, sites from among which the most suitable will be chosen. "Area" here means a part of the territory of a size usually between some tens and some hundreds of square kilometers and which potentially contains suitable sites. The site, which must be able to hold both the plant and its service and restricted access areas, is a part of the territory whose size may be between a few tenths of square kilometers and some square kilometers.

The preliminary selection stage is often called the "site survey stage" and the later one the "site evaluation stage".

Selection of the territorial characteristics to be used in the site survey stage is normally based on their importance at the region of interest level, on the ready availability of the data needed, and on the ease of applying simple exclusion criteria. Examples of such characteristics include population density, surface geological faults, volcanic phenomena, historical seismicity, availability of cooling water, ground slope, and transport infrastructure. In all cases selection is on the basis of reference thresholds for each characteristic examined.

At times, especially when the investigations needed to check the suitability of a site are very costly, and this is always the case when significant *in situ* studies are required, it may be advisable to follow the first preliminary selection with a second based on deeper knowledge which makes it possible to rank the areas still under examination and choose the most promising for siting purposes.

The next stage is acceptance of these areas and identification within them of suitable sites which meet siting objectives. In this stage all the territorial parameters concerned are taken into account and each is studied to the necessary depth, examining all implications for the design, construction, and operation of the plant imposed by each individual site; alternative site-plant relationships are developed if necessary. The final choice of site will be made from among those judged suitable and with a view to optimizing plant-site interactions considered as a whole and with each interaction given its own weight.

D. Significant Territorial Parameters

The aim of this examination is to provide some examples of some territorial parameters which can be used in a preliminary selection process. These same parameters can be used again, but with greater detail and information content, in the site evaluation stage.

1. Clinometry

The clinometry parameter can be examined right from the start of the siting process, in relation to the characteristics of the region of interest, especially when the latter includes a high percentage of mountainous or hilly land. The reference criteria is the maximum acceptable slope for plant construction.

These parameters and others like them, e.g. the difference in height between the plant and a source of cooling water, may lead to exclusion of sizable parts of the region of interest and thus a drastic reduction in the territory to be further examined. They are therefore used as screening parameters in the area identification stage.

2. Territory

The main aim of the investigations of the territory is identification of all the elements which, because of their importance for the community, might be thought decisive for the possibility of siting the plant, e.g., restrictions on land use (scenic, archaeological, monuments, military, airport, hydrological, etc.) and nature reserves and parks, activities potentially harmful to plant safety if carried out in its vicinity, and transport infrastructure (also to be considered in terms of transporting large components to the site). These aspects set limits to site identification. Preliminary information on them is readily available and they are therefore used as principal selection factors for the territory under examination.

3. Demography

Even if the main line of defense for the public lies in the intrinsic safety of the plant, it is always the practice to limit the number of people exposed to risk as far as reasonably achievable. For this reason site selection is directed towards sites which are sufficiently remote from residential settlements and the demographic parameter is an important one in identification of the most suitable sites. It must obviously be considered in relation to the effects it is desired to minimize.

In the case of plants where the greatest hazard comes from explosions or fires, the study area may extend to the greatest distance at which the effects could be great enough to induce lesions, a distance which is at the most of the order of a few kilometers. In the case of the release of toxic or flammable substances heavier than air, the study area might have to be extended to more than ten kilometers, depending on the size of the release, the local geography etc. In the case of a release into the atmosphere of polluting substances which would make their effects felt at great distances from the site of release, the study area may even be extended to the order of hundreds of kilometers; however, at such distances the effects are usually independent of the specific site chosen. In the event of a release of polluting liquid, the study area may extend some tens of kilometers from the site of the release, depending on the surface and underground hydrology of the area.

In evaluating the demographic parameter, account must be taken not only of the number people potentially involved, but also of their location in terms of distance and direction from the plant, since this affects the individual risk to which they are exposed. Normally, people living in the immediate vicinity of the plant have a higher individual risk and their weight in the estimate of overall social risk is therefore higher. Weighting factors for demographic parameters to take account of that aspect may sometimes be necessary.

4. Geology, Seismology and Soil Mechanics

The principal aim of the studies to be conducted on these territorial parameters in plant siting is to acquire all the information needed to estimate whether the site is stable, geologically and in terms of foundation soil, and also whether the plant can be built within the framework of certain predetermined technological solutions. Obviously, these evaluations must be made taking account of the earthquake hazard in the area.

The first aspect, stability, is one of the factors which contribute directly to the decision on the stability of a site for a plant. The second aspect, feasibility — in both the technical and economic sense — may contribute directly to the decision on the suitability of a site, but in any case influences the ranking of the sites which have passed the suitability test. In fact, ensuring the soundness and functionality of structures, systems, and components of importance for plant safety purposes, even in the face of the seismic stresses possible at the site, inevitably involves costs which may differ for the various sites available and which are therefore important in ranking these.

The evaluations required are based on a complex of information obtained from specific studies and investigations supported by a whole series of *in situ* investigations such as site surveys, drillings, tests, installation of recording systems, and laboratory tests. Since *in situ* investigations are usually very costly, it is advisable to make a peliminary selection of the areas of interest through a preliminary examination of the information available on their seismotectonic and geotechnical characteristics. One useful selection tool in this connection is consideration of maximum historical earthquakes, analysis of which is easier and more meaningful the longer the historical period available and the greater the extent to which the data cover the region of interest are. As a threshold value, one can take the maximum earthquake intensity against which defense is economically or technically possible. Another screening tool might be consideration of geological faults, even those not known to be active. However, the criterion of excluding areas near faults should be adopted only for faults already known to be active, although this is a rare case, so as to avoid excluding parts of the territory which might prove suitable. On the other hand, excluding all the above areas would be usefully applied to a region with a wide choice of siting possibilities where seismotectonic knowledge allows for sufficient analysis, thus avoiding a considerable mass of studies and investigations in the faults activity.

5. Hydrology

The aim of hydrological investigations are many. In the first place, there must be sufficient water for the plant to operate (cooling water, washing of raw materials and products, etc.). It is then necessary to evaluate the impact of releases of dangerous and polluting substances on the ground and on water bodies as well as the effect of any possible floods on plant safety. The first aspect is a factor of primary importance in limiting the siting of plants which need large amounts of water to operate. For plants of this type, studies on the availability of water resources (coastal, water courses, lakes) are a priority and a preliminary selection of areas is made on the basis of the results.

The second aspect is not usually a factor in preliminary selection.

The third aspect may be a precluding factor if it turns out that there is a possibility of flooding by landslides or dam failures. This point must be examined right from the start of the siting process. The effects of overflowing rivers can be prevented by appropriate defensive works (levees, barriers, etc.), but the requirements the site imposes on the plant design will contribute to its ranking in final site selection.

6. Meteorology

Data and information on the meteorological characteristics of the area must be collected in order to evaluate the impact on man and on the environment of releases of dangerous or simply polluting substances into the atmosphere, as well as to evaluate the influence of exceptional meteorological phenomena on plant safety.

It is not likely that these aspects would play a role in area selection unless it was possible to ascertain the occurence of exceptional phenomena of extreme severity which would probably be a reason for rejection of such an area. On the other hand, these parameters are of importance in the identification of suitable sites and in ranking them. Usually, from this point of view, the normal spread of pollutants which determines the magnitude of impacts is more important than the occurence of exceptional meteorological phenomena. These latter, except in special cases where significant additions must be made to plant design, do not generally cause any difficulty.

III. EMERGENCY PLANNING

By emergency we mean a sudden, generally unexpected, occurrence or set of circumstances demanding immediate action. By emergency planning, we mean the previous planning of such actions.

On the question of major hazards plants, it should be noted that a series of actions to be taken promptly following an accident or a potentially dangerous situation can contribute effectively to mitigating the possible adverse consequences of such an accident or situation, particularly in terms of health detriment. This is not only because the safeguarding of public health is a priority objective in the protection field but also because such actions are normally less effective in reducing the nonhealth consequences. However, if the actions taken are to be effective they must be thoroughly planned in advance. It is obvious that their effectiveness will be the greater, the shorter the time needed to put them into effect is and only when a plan is already available it is possible to prevent the delays otherwise unavoidable in emergency situations, just the time when speed of action becomes of fundamental importance. It is equally obvious that an effective reduction of the harm resulting from an accident cannot be ensured by merely planning actions, however good the plan may be. A scrupulous commitment to implementation, going beyond the contents of the plan itself, must also be ensured.

We shall certainly not dwell here on the problems of implementing emergency plans. Rather we shall discuss the more important aspects of emergency planning, especially that external to the plant, and the reference criteria for such planning

First, we should make it clear that the distinction which is usually made between planning for internal (on site) emergencies and that for external (off site) emergencies, where the adjectives internal and external refer to the plant area, has its origin in the difference in the main objectives of the two types of planning: containing the accident in the first case and containing the consequences in the second.

On the subject of internal emergency planning, attention should be drawn to the fact that containment of an accident situation is achieved — as far as possible — firstly and principally by control of and action on the source of the accident, i.e., the plant. This requires the taking of a certain number of actions (direct intervention on the plant, measuring the course of the accident, measuring the amount and nature of dangerous substances released by the plant, etc.) which must be carried out within the plant or in its immediate vicinity. These actions are the subject of the plan.

Having an area around the plant where such actions can be taken free from interference at times when taking them is an absolute priority is an essential prerequisite for the success of any emergency action. This area is normally called exclusion area and is usually an integral part of the plant itself. In addition to being a most valuable feature for the management of any sort of emergency, it serves to isolate the activities of the surrounding area from those of the plant, thus limiting both the external impact of any accident and also any adverse effects of the first activities on the safety of the second. For this reason, too, the adoption of an exclusion area around major hazards installations is highly advisable.

For the actions required in an emergency to be carried out effectively, the exclusion area must be suitably prepared. For example, there must be more than one emergency entrance to the exclusion area and these must be situated in different directions, so that operations are not forced to be from down wind. There must also be alternative emergency control posts situated on the periphery of the exclusion area and accessible during an accident. There must also be roads to permit rapid execution of the emergency actions envisaged, both within the exclusion area and outside it.

Turning to external emergency planning, given the multiplicity of the possible impacts, it is necessary to define priority objectives, the more so when the accident may be a major one and therefore require mobilization of considerable resources, perhaps for a long period. It is therefore necessary to establish an order of precedence to achieve the best use of men and materials. The objectives can be formulated as follows.

1.   To save endangered human lives and and aid those seriously affected.
2.   To provide shelter and aid for those who have to leave their homes.
3.   To ban the use of contaminated foodstuffs, when necessary, and aid those affected by the ban.
4.   To ensure other essential social services.

While these objectives are common to all emergency situations, the measures to be taken will depend directly on the nature of the possible accidents to the plant (type of substances released, form of release, amount of release) and therefore on the type of plant under consideration. Consequently, the measures taken may vary widely. As examples, one can think either of a nuclear installation or of a processing plant with flammable or toxic substances involved in a very serious accident.

In the first case, the duration of a release goes from orders of magnitude of 1 h to some days. The harmfulness of the substances released persists over time (from some days to some years according to the substance) and is related mainly to their inhalation or ingestion. The area involved in the release may be very extensive (greater involvement near the point of release gradually decreasing with distance) and there may be heavy contamination of the ground near the release point (even over some kilometers). The population may be wholly or partially evacuated for up to some tens of kilometers downwind from the release and there may be controls over foodstuffs for up to a hundred kilometers. Intervention may extend over periods of weeks or months.

In the second case, for example, a release of substances heavier than air (chlorine or LPG) into the atmosphere, the duration of the release is between a few minutes and a few hours. Danger from the release, except in the case of improbable local build-ups, is linked to inhalation (in the case of chlorine) or to fire or explosion (in the case of LPG) and may extend for a few kilometers (LPG) or a few tens of kilometers (chlorine) from the point of release. Ground contamination is also limited or nil. Evacuation is short term and intervention is also limited in duration (a few days).

It is clear at this point that the types of action to be planned and the methods for their implementation will be markedly different in the two cases. This difference should not, however, make one think of separate planning and separate organizations for implementation. In spite of the differences mentioned, the majority of the actions will be taken, at national, regional, and local levels, by the same organizations — e.g., the health, police, and civil protection organizations — and many of the means employed can adequately meet differing needs. Optimization of available resources and coordination of those involved requires all specific plans to be included within a general emergency plan for the area.

The basic technical components of any plan must derive from a complete analysis of the full spectrum of possible accidents, even those usually considered improbable, and from a thorough examination of the territory surrounding each plant. These will make it possible to identify a range of possible scenarios for the external situation and to evaluate the possible consequences of accidents in the absence of intervention.

The methods and tools available for analysis and evaluation of these consequences are described in other parts of this book.

The intention here is rather to make a few general observations.

Major accidents usually have a low probability of occurrence. There is therefore often a temptation not to plan any action for the most serious accidents because the probability of their occurrence is so low and because the scope of the actions required to tackle them makes detailed prior identification difficult if not impossible.

This is not an attitude which can be shared. The need to prepare an emergency plan even for accidents with low probabilities should be appreciated not only for obvious political reasons, since the lack of such a plan could create psychological problems for those members of the public most exposed to the risk, but also because the wider area over which action would be required in the event of a serious accident would necessitate wider involvement and therefore the overlapping of a multiplicity of competences and responsibilities. The resulting situation could become practically ungovernable if not regulated in advance. From this point of view it is not the total risk (probability combined with the extent of the consequences) which should be taken as the reference criteria in emergency planning, but only the component represented by the extent of the consequences. It would therefore be necessary in some cases, given the wide range of possible consequences, to prepare two plans of action, one of a local character for minor accidents with higher probabilities and another on a larger (regional or national) scale for major accidents with remote possibilities.

There may again be a temptation to feel that preparation of two emergency plans is too burdensome, the second requiring considerable resources to be tied up in expectation of an event which will most probably never occur at all. The problem of probability comes up again. It now seems to have some weight in these considerations. The solution is to be sought in detailed planning of the actions to be taken in the event of minor accidents and only more general planning of those for major ones. However, even for these last it is necessary to lay down clearly those features which would be difficult to establish during an emergency, such as the different parties required to act, their tasks and responsibilities, and the tools they can use. More detailed planning would be difficult to implement and not always correct, since the way in which a major accident situation will evolve can be predicted only in outline.

In their more general aspects these considerations apply to all major hazards plants, but more especially to those in which the more serious accidents may have results which require emergency action over wide areas and therefore go far beyond the response capability and possibilities for management of a local emergency plan. On the other hand, when the most serious imaginable accident does not go beyond a local dimension, there is no need for two emergency plans, only for one in which actions to be taken for the most serious accidents are planned in detail.

As far as the territory is concerned, adequate consideration must be given in the drawing up of emergency plans to all those local characteristics which condition the implementation of the measures or the variability of which may result in the failure of cetain emergency measures which it is desired to provide for. It must first be pointed out that, in general, the siting requirements for this type of plant lead to them being located in areas with geographical characteristics facilitating implementation of emergency measures. Special situations, which require provision for special actions, may obviously appear in exceptional cases, e.g., the presence of an inhabited island near an industrial establishment or a plant sited in a narrow valley.

Another aspect to be pointed out results from the fact that since such plants are usually situated near the coast or estuaries or on the banks of rivers, the presence of the sea or river interrupts the continuity of the territory surrounding the plant. Since in all cases one has to work downwind as little as possible, it is necessary to prepare emergency measures which can start from more or less opposite sides of the plant.

A third aspect which requires careful consideration is soil conditions in the area around the plant. These, together with the characteristics of the road network in the same area, may be of decisive importance in preparing adequate emergency plans.

The presence of an extensive road network is a matter of great importance in emergency planning, both for rapid access to the area of the emergency, for speedy evacuation of the area, and in terms of ability to operate from upwind in all situation.

Railway lines near the plant may add to the problems. Still on the question of roads and railways, attention must be given to the presence of tunnels and the possible problems these may cause in the event of serious releases.

Finally, there is climate. The factors of importance are those which could hinder emergency operations, e.g., snow or fog. These must be taken into account, particularly in relation to the frequency of their occurrence in the area, so that provision can be made to tackle them adequately.

One aspect which must never be overlooked in emergency planning is the transition from the period of unquestionable emergency to the subsequent postemergency period. The latter may be extremely costly in terms of men and materials and may last very much longer than the emergency proper in the case of major accidents which require, for example, complete or partial evacuation of wide areas surrounding the plant and health checks on large segments of the population. This period is also of great importance in reducing the adverse consequences of an accident, and even if it is difficult to describe it as strictly emergency it should nevertheless be foreseen and, at least in outline, planned and correlated without any discontinuity with the preceding emergency plan. Otherwise there is a risk of finding a very harmful gap in planning after the most essential urgent action has been taken.

REFERENCES

1. *Siting of Nuclear Facilities,* Pro. Ser. STI/PUB/384, International Atomic Energy Agency, Vienna, 1975.
2. *Safety in Nuclear Power Plant Siting,* A Code of Practice, Safety Series No. 50-C-S, International Atomic Energy Agency, Vienna, 1978.
3. *Canvey — An Investigation of Potential Hazards from Operations in the Canvey Island/Thurrock Area,* Health and Safety Executive, London, 1978.
4. **Kunreuther, H. C., Linnerooth, J., Lathrop, J., Atz, H., Macgill, S., Mandl, C., Schwarz, M., and Thompson, M.,** *Risk Analysis and Decision Processes — The Siting of Liquefied Energy Gas Facilities in Four Countries,* Springer-Verlag, Berlin, 1983.
5. *Emergency Planning and Preparedness for Nuclear Facilities,* Proc. Ser. STI/PUB/701, International Atomic Energy Agency, Vienna, 1986.
6. *Planning Emergency Response Systems for Chemical Accidents,* Interim Document, Regional Office for Europe, Copenhagen, World Health Organization, 1981.

Index

INDEX

Printed and bound by CPI Group (UK) Ltd, Croydon, CR0 4YY

17/10/2024

01775700-0002